Handbook of

Nanofibers and Nanocomposites

Handbook of

Nanofibers and Nanocomposites

Characteristics, Synthesis, and Applications in Textiles

edited by

Mohd Yusuf

Aminoddin Haji

JENNY STANFORD
PUBLISHING

Published by

Jenny Stanford Publishing Pte. Ltd.
101 Thomson Road
#06-01, United Square
Singapore 307591

Email: editorial@jennystanford.com
Web: www.jennystanford.com

British Library Cataloguing-in-Publication Data
A catalogue record for this book is available from the British Library.

Handbook of Nanofibers and Nanocomposites: Characteristics, Synthesis, and Applications in Textiles

ISBN 978-981-4968-77-5 (Hardcover)
ISBN 978-1-003-43274-6 (eBook)

Contents

Preface xi

1. Polymer-Based Nanofibers **1**

Adeleh Gholipour-Kanani and Niloofar Eslahi

1.1 An Introduction to Nanoscale Materials 2
1.2 Nanofibers as Promising Materials 4
1.3 Fabrication Methods of Nanofibers 4
1.3.1 Phase Separation 5
1.3.2 Drawing 7
1.3.3 Template Synthesis 7
1.3.4 Self-Assembly 8
1.3.5 Centrifugal Spinning 9
1.3.6 Blown Bubble-Spinning 10
1.3.7 Electrospinning 11
1.3.8 Electro-centrifugal Spinning 12
1.4 Different Types of Nanofibrous Structures 14
1.4.1 Randomly Oriented Nanofibers 14
1.4.2 Aligned Nanofibers 15
1.4.3 Porous Nanofibers 16
1.4.4 Hollow Nanofibers 18
1.5 Nanofibers for Technical Textiles 19
1.5.1 Filtration 19
1.5.2 Automotive Applications 22
1.5.3 Geotextiles 25
1.5.4 Medical Textiles 27
1.5.4.1 Wound dressing 27
1.5.4.2 Tissue-engineered scaffolds 30
1.5.4.3 Drug delivery 33
1.6 Conclusion and Future Perspectives 35

2. Science and Applications of Polymeric Nanofibers **47**

S. Chakraborty, Pragati Bajpai, L. Chakraborty, and Shanu Prabhakar

2.1 Introduction 48

2.2 Polymeric Materials for Nanofibers 48
2.3 Production of Nanofibers: Various Techniques 49
2.3.1 Drawing 49
2.3.2 Spinneret-Based Tunable Engineered Parameters 51
2.3.3 Template Synthesis Method of Producing Nanofibers 51
2.3.4 Thermally Induced Phase Separation 53
2.3.5 Polymer Nanofibers by Molecular Self-Assembly 55
2.3.6 Electroconducting Nanofibers: Interfacial Polymerization 57
2.3.7 Electrospinning of Nanofibers 57
2.3.7.1 Components of the electrospinning process 58
2.3.7.2 Different types of nozzles 58
2.3.7.3 Needleless spinneret 60
2.3.7.4 Electrospinning setups and different collector types 60
2.4 Material and Process Variables in Electrospinning 60
2.4.1 Material and Process Parameters Affecting Electrospinning 60
2.4.1.1 Electrospinning: Effect of material variables 61
2.4.2 Effect of Process Variables 62
2.4.3 Electrospinning Technology: Mathematical Modeling 62
2.4.3.1 Bending instability 65
2.5 Applications of Polymeric Nanofibers 70
2.5.1 Filtration 70
2.5.2 Biomedical Applications 72
2.5.2.1 Tissue engineering 72
2.5.2.2 Drug delivery 73
2.5.2.3 Wound dressing 73
2.5.3 Applications in Protective Clothing 74
2.5.4 Applications in Energy Storage 74
2.5.5 Applications as Nanocomposites 75
2.5.6 Environmental Application 76
2.5.7 Applications in Sensors 77
2.6 Conclusion 77

3. Nanoyarns: Recent Advancements in Production Techniques, Applications, and Future Prospects **87**

Çağlar Sivri

3.1 Introduction 87
3.2 Nanofibers' Timeline and Yarn Production 88
3.2.1 Properties of Nanoyarns 90
3.3 Nanoyarn Production Techniques 90
3.3.1 Nanoyarns via Electrospinning 91
3.3.2 Nanoyarns via Centrifugal Spinning 100
3.3.3 Nanoyarns via Solution Blowing (Air-Blowing) 102
3.4 Applications of Nanoyarns 106
3.4.1 Nanoyarns in Tissue Engineering 106
3.4.2 Nanoyarns in Drug Delivery 108
3.4.3 Nanoyarns in Energy 109
3.4.4 Nanoyarns as Sensors 109
3.4.5 Future Potential Applications 111
3.5 Future Prospects in Nanoyarns 111
3.6 Conclusion 115

4. Drug-Loaded Nanofibers: Production Techniques and Release Behaviors **125**

Hülya Kesici Güler and Funda Cengiz Çallioğlu

4.1 Introduction 126
4.2 Production Techniques 127
4.2.1 Surface Modification 128
4.2.2 Blend Electrospinning 129
4.2.3 Suspension Electrospinning 131
4.2.4 Coaxial Electrospinning 134
4.2.5 Emulsion Electrospinning 135
4.2.6 Microcapsule Loaded Nanofibers 139
4.3 Conclusion 142

5. Textile Applications of Nanofibers and Nanocomposites **149**

Shumaila Kiran, Shahid Adeel, Shazia Abrar, Sarosh Iqbal, Saba Naz, and Nimra Amin

5.1 Introduction 150
5.2 Nanomaterials in Textile Industry 151

5.3 Nanotechnology Applications in the Textile Industry 152
5.3.1 Nanofinishing 152
5.3.2 Nanocoating 153
5.3.3 Nanocomposites 154
5.4 Functional Nanomaterial Textiles 155
5.4.1 UV Protective Textiles 155
5.4.2 Flame-Retardant Textiles 156
5.4.3 Water- and Oil-Repellent Textiles 156
5.4.4 Antimicrobial Textiles 157
5.4.5 Wrinkle-Resistant Fabrics 159
5.5 Future Prospects 159
5.6 Conclusions 159

6. Nanomaterials in Textiles: Performance, Health, and Environmental Aspects 169

Liliana Indrie, Sabina Gherghel, and Steven McNeil

6.1 Introduction 170
6.2 Nanotechnology in Textiles 171
6.3 Nanoparticles 172
6.3.1 Routes to Incorporating Nanoparticles into Textiles 172
6.3.1.1 Application of prepared nanoparticles 172
6.3.1.2 *In situ* preparation 173
6.3.2 The Effect of Nanoparticles on the Handle of Textiles 174
6.4 Applications of Nanoparticles in Textiles 175
6.4.1 Nanoparticle-Textile Catalysts 175
6.4.2 Protecting Textiles from Insect Damage 176
6.4.3 Resistance to Dry Soiling 176
6.4.4 Self-Cleaning Applications 178
6.4.5 UV Protection 179
6.4.6 Antimicrobial Applications 180
6.4.7 Other Textile Properties 180
6.5 Potential Effects of Nanomaterials on Health and the Environment 181
6.6 Conclusions 183

7. Overview of Polymer/Metal-Oxide Nanocomposites: Synthesis, Properties, and Their Potential Applications **193**

Bilal Ahmed, Arvind Singh, Ruhinaz Ushal, Mohd Yusuf, Sachin Kumar, Animesh Kumar Ojha, and Wasim Khan

7.1 Introduction 194
7.2 Synthetic Techniques 196
7.2.1 Physical Vapor Deposition 196
7.2.2 Thermal Evaporation 196
7.2.3 Pulsed Laser Deposition 196
7.2.4 Chemical Vapor Deposition (CVD) 197
7.2.5 Sol-Gel Technique 198
7.2.6 Co-precipitation Techniques 198
7.2.7 Solvothermal Techniques 198
7.2.8 *Ex/In situ* Formation 199
7.2.8.1 *Ex situ* process 199
7.2.8.2 *In situ* process 199
7.2.9 Processing of Polymer/Metal-Oxide Nanocomposites 199
7.3 Properties of Polymer/Metal Oxide–Based Nanocomposites 200
7.3.1 Physical, Mechanical, and Rheological Properties 201
7.3.2 Thermal and Chemical Properties 201
7.3.3 Electrical and Optical Properties 202
7.3.4 Biological Properties 202
7.4 Applications 203
7.5 Conclusions and Future Trends 205

8. Self-Cleaning Nanofinishes and Applications **221**

Akhtarul Islam Amjad

8.1 Introduction 222
8.2 Inspiration for Self-Cleaning Surface and Historical Development 223
8.3 Basic Mechanisms of Self-Cleaning 225
8.4 Methodology for Developing the Self-Cleaning Textile Material 226
8.4.1 Self-Cleaning Textile Finishing by a Hydrophobic Material 227

8.4.1.1 Plasma technology 227
8.4.1.2 Electrospinning technology 228
8.4.1.3 Sol-gel treatment 229
8.4.1.4 Fluorinated polymers 231
8.4.2 Self-Cleaning Textile Finishing by Hydrophilic/Hydrophobic Nanomaterial 232
8.4.2.1 Photocatalyst 232
8.4.2.2 Microwaves 235
8.4.2.3 Carbon nanotubes 236
8.4.2.4 Silver nanoparticles 236
8.4.2.5 Colloidal metal oxide 237
8.4.2.6 N-halamine 237
8.5 Self-Cleaning Finishes on Different Textile Substrates 238
8.5.1 Self-Cleaning Cotton Fibers/Fabrics 238
8.5.2 Self-Cleaning of Protein Fibers/Fabrics 239
8.5.3 Self-Cleaning Polyester Fibers/Fabrics 240
8.5.4 Self-Cleaning Modified Cellulose Fibers/Fabrics 241
8.6 Evaluation of Self-Cleaning Textiles 241
8.7 Applications of Self-Cleaning Textiles 242
8.8 Limitations of Self-Cleaning Fabric 243
8.9 Conclusion 244

Index 257

Preface

In recent years, nanotechnology has been playing an important role in the development of various industries including the textile sector. Textiles with functional properties such as antimicrobial finishes, drug delivery, ultraviolet resistance, electrical conductivity, superhydrophilicity, superhydrophobicity, self-cleaning, EMI shielding, flame-retardance, etc. can be produced with the help of nanotechnology. Nanomaterials can be added to textile materials at different stages of the production process, including spinning, finishing, and coating. Nanocomposites are a result of the combination of nanoparticles and ordinary textile substrates and show enhanced properties based on the applied nanoparticle and processing. Nanomaterials can be embedded in the core, sheath, or whole bulk of the polymeric matrix of the fibers. They can also be applied on the surface of the textile materials in the form of coatings with various compositions or through application methods such as spraying, laminating, printing, etc.

Nanofibers are textile materials that show enhanced properties due to larger surface area compared with ordinary textile fibers. They have diameters lower than 1000 nm and can hold nanoparticles, drugs, extracts, essential oils, etc. in their polymeric matrix. They actually encapsulate these compounds and are able to control their release by delivering them only at the targeted sites. Polymeric nanofibers have found various applications due to their distinctive features such as large surface area, high porosity, and superior chemical and physical properties. Recently, nanofibers and textile nanocomposites have attracted great interest in the industry and research and electrospinning is the most famous method among the several methods that have been developed for the fabrication of nanofibers.

This book is a collection of the reviews on the recent advances in the fields of nanofibers, nanocomposites, and their applications in textiles as well as related fields. Chapters 1 and 2 introduce nanomaterials and the different types of nanofibers. The chapters discuss the various methods for fabricating nanofibers and summarize their applications as technical textiles, geotextiles, medical textiles, etc.

Chapter 3 comprehensively overviews the nanoyarn production by the electrospinning technique as well as other manufacturing possibilities such as solution (air) blow spinning and centrifugal spinning. The future prospects and insights into nanoyarn production are also discussed in this chapter. Chapter 4 presents the various techniques for the fabrication of drug-loaded nanofibers, including blend, suspension, emulsion, coaxial electrospinning, etc. and compares them with respect to their release behaviors, time, and properties.

Chapters 5 and 6 review the various applications of nanotechnology in the textile industry, such as UV protection, flame retardancy, water and oil repellency, antimicrobial finishing, and wrinkle resistance in textiles, and the potential environmental and health aspects related to them. Chapter 7 presents a brief review of nanocomposites and bio-nanocomposites and their textile applications. Chapter 8 discusses the self-cleaning property of textiles, which is an important application of nanotechnology.

The authors who contributed to this book are specialists in field of nanotechnology, nanocomposites, and nanofibers. The editors therefore hope that the students, researchers, and academicians from various fields, such as textile finishing, medical textiles, chemical engineering, and materials science will find this book of great interest and useful in their curriculum.

Finally, we sincerely thank those who supported this book in anyway. We acknowledge the great efforts of the eminent authors without whom this book would have been unimaginable. We also appreciate the interest shown and the support given by the publisher, which allowed us to bring out this handbook.

Mohd Yusuf

Aminoddin Haji

March 2023

Chapter 1

Polymer-Based Nanofibers

Adeleh Gholipour-Kanani and Niloofar Eslahi
Department of Textile Engineering, Science and Research Branch, Islamic Azad University, 1477893855, Tehran, Iran
a.gholipour@srbiau.ac.ir, niloofar.eslahi@srbiau.ac.ir

Nanofibers are one of the most applicable nanostructures with an average diameter of less than 1 μm which can be made from various kinds of raw materials such as ceramics, polymers, metals, and so on. In this context, polymers are of more interest due to their higher processability, more variety, and the possibility of creating desired properties for specific performance. Polymeric nanofibers have been implemented in different applications owing to the distinctive features of nanofibers, i.e., large surface area, high porosity, as well as superior chemical and physical properties. Nanofibers made from synthetic polymers often exhibit good physical and mechanical properties, while nanofibers made from natural polymers are best known for their excellent biological properties useful for biomedical applications. In this regard, hybrid nanofibers, with the aim of creating desirable physical and mechanical properties along with suitable biocompatibility and biodegradability, have been highly studied by

Handbook of Nanofibers and Nanocomposites: Characteristics, Synthesis, and Applications in Textiles
Edited by Mohd Yusuf and Aminoddin Haji

ISBN 978-981-4968-77-5 (Hardcover), 978-1-003-43274-6 (eBook)
www.jennystanford.com

scientists. Several strategies to produce polymer nanofibers have been identified including phase separation, drawing, template synthesis, self-assembly, conventional electrospinning, centrifugal spinning, etc. Due to the appropriate features of the electrospinning approach such as cost-effectiveness, ease of production, the ability to produce continuous fibers in controllable and wide diameter ranges, and so on, most of the techniques have been modified based on this method to achieve mass production such as electro-centrifugal spinning. In this chapter, different aspects of polymer nanofibers from different fabrication methods, parameters affecting the morphology and patterns of nanofibers, and their various applications in different industries will be studied.

1.1 An Introduction to Nanoscale Materials

Nowadays, nanoscale or nanostructure materials, which are also called nanomaterials, have attracted enormous interest in science and industry. According to the definition of nanotechnology by the National Nanotechnology Initiative (NNI), nanoscale materials are special structures that have at least one dimension below 100 nm. These kinds of materials have a special characteristic that influences and determines their all properties and that is their high specific surface area [1]. This fact means more surface area is accessible in a lower mass of materials. Therefore, due to the availability of a much greater number of functional groups, the performances of nanomaterials are considerably higher than their counterparts on larger scales. For example, a material such as titanium dioxide (TiO_2), which is an inexpensive biocompatible semiconductor, has unique characteristics due to its crystalline structure, including UV light resistance, high chemical stability, and good photocatalytic property that lead to its fungicides and germicidal properties [2]. In this case, a certain amount of TiO_2 compared to the same amount of TiO_2 nanoparticles have less crystals on the surface and most of them are inaccessible and inactive in the bulk. Therefore, even in small quantities, nanoscale materials have properties several hundred times greater than the same materials on larger scales.

Nanoscale materials are classified into four groups according to the number(s) of their dimensions larger than 100 nm [3]. Zero-dimensional (0D) nanomaterials are structures whose all of their dimensions are in the nanoscale range. Nanoparticles, nanospheres, and quantum dots are some examples of this class [4]. One-dimensional (1D) nanomaterials such as nanofibers, nanowires, nanotubes, and nanorods which have only one dimension outside the nanoscale range, have attracted extensive attention owing to their unique characteristics [5]. Two-dimensional (2D) nanostructures have only one dimension in the nanoscale range and the other ones are outside the nanometer range. Nanoplates [6], nanodisks [7], nanosheets [8], nanobranched clusters [9], and so on are examples of 2D nanostructures. Three-dimensional (3D) nanomaterials are materials that are not confined to the nanoscale in any dimension. This type of nanostructure has attracted significant attention as the building block of high-performance nanodevices [4]. Some kinds of 3D nanostructures are dendritic structures such as nanoballs [10], nanocoils [11], nanocones [12], nanopillars [13], and nanoflowers [14]. Figure 1.1 represents some examples of different kinds of nanoscaled materials.

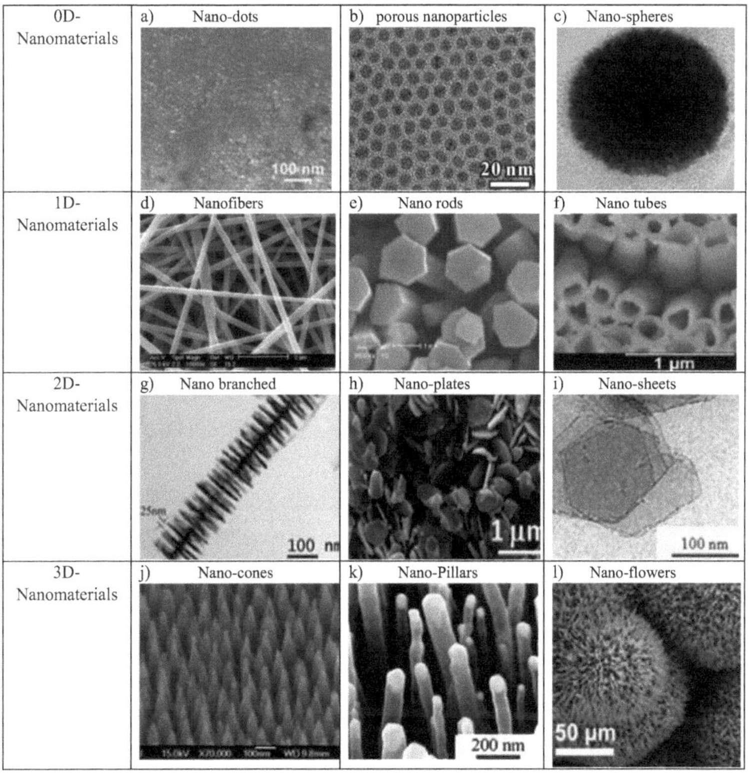

Figure 1.1 Examples of different kinds of 0D, 1D, 2D, and 3D nanoscale materials (a, c, e–l [4], b [1], and d [15]).

1.2 Nanofibers as Promising Materials

Fibers are flexible and fine substances with a high ratio of length to a diameter which are made of natural or synthetic materials including polymers, metals, ceramics, etc. Conventional fibers have a diameter in the range of 5 to 50 μm and are widely used in different industries. During the past decades, with progress in nanotechnology, nanofibers and nanofibrous structures have attracted great attention. Nanofibers were among the first nanostructures to be identified and interested due to their high processability and applicability. Although the fibers with a diameter below 100 nm were called nanofibers in their initial definition, in recent years, all fibers with less than one micrometer in diameter and an aspect ratio (length to diameter ratio) of at least 100:1 are considered nanofibers [16].

Because of the flexibility and continuous form of nanofibers, they have been considered promising nanostructures in different applications. The applications of nanofibers have improved in various fields such as biomedicine, IT, energy saving and environment cleanup, life science, and so on. Some of the main features of nanofibers are their high surface-to-volume ratio, great mechanical and thermal properties, and special assemblies which leads to 3D and porous structure with excellent pore interconnectivity [17, 18]. Combining nanofibers' unique properties with the special materials' functionalities would result in novel properties and applications. Polymers are of more interest materials in nanofibers fabrications due to their higher process-ability, more variety, and the possibility of creating desired functionality for specific performance. Polymer-based nanofibers are considered for many applications in different industrial features such as filtration [19], (bio)chemical sensors [20], cosmetic skin masks [21], nanomedicine [22], tissue engineering [23], wound dressing [24], drug delivery [25], electronic devices [26], etc.

1.3 Fabrication Methods of Nanofibers

The methods for fabricating nanofibers can be divided into two main groups: (1) non-electrospinning methods by using mechanical force including drawing, phase separation, self-assembly, template synthesis, etc., and (2) electrospinning methods by using electrostatic force [27, 28]. Different setups for the fabrication of nanofibers are shown in Fig. 1.2. Each method is explained comprehensively in this section.

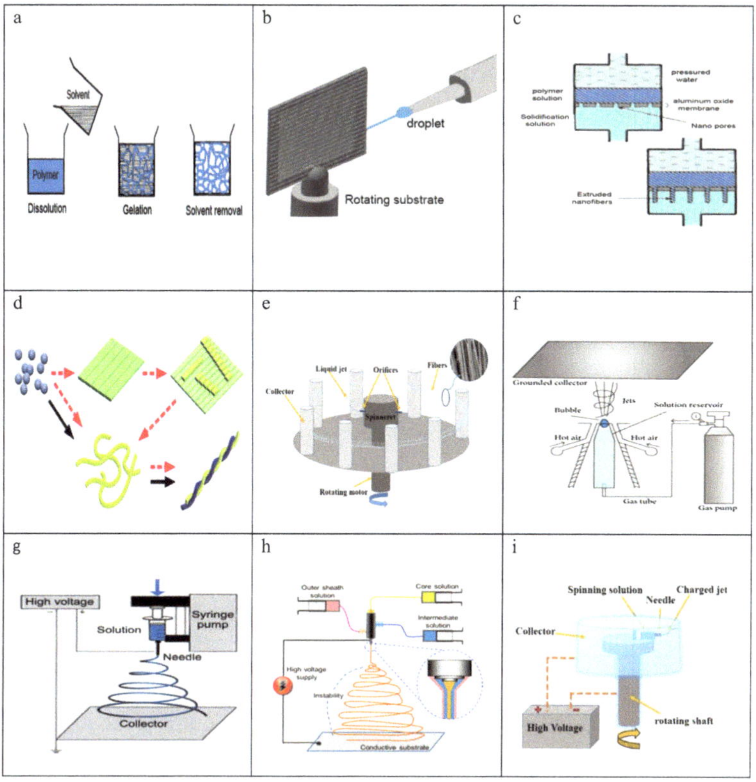

Figure 1.2 Schematic illustration of (a) phase separation [27], (b) drawing [29], (c) template synthesis [27], (d) self-assembly [41], (e) centrifugal spinning [68], (f) blown bubble spinning [29], (g) conventional electrospinning [29], (h) triaxial electrospinning [29], and (i) electro-centrifugal spinning [68] setups.

1.3.1 Phase Separation

This technique is based on the physical incompatibility of at least two phases. For this purpose, at first, the polymer dissolves in a solvent followed by gelation. The solvent phase is then extracted from the solution, but the other phase remains (Fig. 1.2a). Freeze/drying is performed under a vacuum to remove the solvent and a porous nanofiber structure is formed [27]. The effective parameters are the freeze/drying temperatures, the concentration of polymer, the selection of proper solvent, and the presence of cross-linking agents [29]. However, only a few polymers with gelation capability, including poly(L-lactic acid), poly(hydroxyalkylmethacrylate)-graft-poly(L-lactide), polyglycolide, polyhydroxyalkanoates, chitosan, and gelatin, have been used in phase separation method for production of nanofibers [30–33]. It should be noted that long continuous fibers cannot be produced by using this method [27].

Thermally induced phase separation (TIPS) arises from a thermodynamically unstable polymer solution for producing composite nanofibers with porous architecture through phase separation processes [34]. The preparation steps include dissolving a polymer in a suitable solvent, phase separation, solvent exchange, and freeze-drying [31]. TIPS is classified into two main methods of liquid–liquid and solid–liquid phase separation. The former method occurs in a polymer solution with an upper critical temperature, whereas the latter method involves solvent crystallization from a polymer solution [30]. Zhao et al. [30] produced chitosan acetate nanofibers via a solid–liquid phase separation method. According to the obtained results, phase separation temperature as well as the concentration of chitosan and acetic acid affected the morphology of the produced fibrous structure. In another study, Liu et al. [31] fabricated 3D nanofibrous gelatin scaffolds via combined TIPS with a porogen-leaching technique for tissue engineering applications. The introduced method could control macropore shape and size by paraffin spheres, pore morphology by phase separation parameters, as well as mechanical properties by polymer concentration and cross-linking density. Samitsu et al. [35] also reported the production of a mesoporous network structure via a phase separation process from frozen polymer solutions with a high extraction rate. In fact, the crystallization of solvent below its freezing point led to ultrafine microphase separation and the formation of an interconnected polymer nanofiber network.

1.3.2 Drawing

The drawing method is similar to dry spinning for the production of nanofibers. A sharp tip or a micropipette is employed to draw a droplet of a previously deposited polymer solution (Fig. 1.2b). After solvent evaporation, the liquid fibers solidified to form single nanofibers owing to the large surface area [27]. The processing parameters including drawing speed and polymer concentration can be tailored to adjust the diameter of the fabricated fibers. The merit of this technique is the possibility of studying the properties of individual nanofibers. The produced single nanofibers can be used in tissue engineering, nanoelectronics, nano-optics, and the production of yarns [36]. Despite the simplicity of the process, there are some limitations such as low productivity, fibers dimension, and materials selection (only possible for viscoelastic polymers). The major disadvantage of this method is that the drawing time is limited. In fact, the droplet viscosity steadily increases with solvent evaporation and the droplet volume declines over time, leading to a change in the fiber diameter. In order to resolve this issue, hollow glass micropipettes are employed to keep the amount of polymer constant [29]. Ma et al. [37] fabricated polyethylene nanofibers using the drawing method from a polymer solution with enhanced thermal conductivity. Recently, Yadavalli et al. [38] employed the novel method of gravity fiber drawing to produce single filament nanofibers from polymer solutions. Therefore, the production of complex 3D scaffolds consisting of fibers with different diameters and organization, controllable spacing, and angular orientation between nanofibers, would be feasible by this method.

1.3.3 Template Synthesis

Another method for producing nanofiber from various materials (polymers, metals, semiconductors, and ceramics) is template synthesis using hard or soft templates. In this process, precursors are first coalesced with templates by impregnation, then solid materials are cast during the reaction, and the final product is achieved after template removal by either chemical methods, including calcination and etching, or physical methods such as dissolution based on the nature of the template [39]. A schematic of template synthesis is shown in Fig. 1.2c. The merit of this method is to control the dimension of the produced nanofibers by using different templates. On the other hand, the length of the nanofibers produced by this method is limited [27]. Usually, the fabricated nanofibers have a hollow structure since the synthesized polymer is subjected to precipitate onto the inner surface of the hollow channels. Tao et al. [40] used template synthesis to fabricate arrays of nanowires and nanofibers from the biodegradable poly(ε-caprolactone) (PCL). This solvent-free and low-temperature fabrication method could be potentially used for the encapsulation of therapeutics for drug delivery.

1.3.4 Self-Assembly

In this bottom-up fabrication technique, the molecules are organized into nanofibers by non-covalent intermolecular interactions including electrostatic reactions, hydrophobic forces, and hydrogen bonding depending on the chemical structures of the molecules [27, 41]. One mechanism involves hydrogel formation from two interpenetrated phases, i.e., the liquid and the solid phases. Dried nanofibers are produced by the self-assembly of hydrogelator molecules after the liquid phase is removed from the hydrogels (Fig. 1.2d) [42]. It is worth mentioning that this method is only limited to biological molecules, i.e., peptides, with a small fiber diameter of less than 20 nm, which can self-assemble by themselves or under external stimuli [43, 44]. As for the disadvantages of this method, its complexity, low speed, low productivity, and limitation in fiber dimension tuning can be considered [27].

Liao et al. [41] studied the mechanism of nanofibers' self-assembly using a small peptide amphiphile. They found that in the solution, peptide amphiphiles nucleate to grow into nanofibers. On the substrates, peptide amphiphiles tend to produce nanofibers and nanosheet rod-like structures concurrently. Chen et al. [45] developed an antimicrobial peptide that self-assembled to form a supramolecular nanofiber with improved membrane selectivity and cytocompatibility. Recently, chitosan and polydopamine were introduced into electrospun silk fibroin nanofibers through layer-by-layer self-assembly to enhance the antibacterial ability and cytocompatibility [46]. Cui et al. [47] also fabricated alginate fibers that encapsulated in a self-assembly peptide hydrogel by microfluidics for wound healing applications. The dual-drug delivery system showed sustained release profiles in which the release of antibiotics from the peptide hydrogel is rapid, but the growth factor release occurred within a week.

1.3.5 Centrifugal Spinning

In rotary spinning, centrifugal spinning, or force spinning, fibers are fabricated by centrifugal forces from a wide range of materials at high speed and low cost. The centrifugal spinning setup is composed of an electric motor-driven rotating spinneret coupled to a polymer reservoir and a collector inspiring the cotton candy production principle [48]. The polymer solutions or melts are fed from a syringe pump or an extruder into a rotating chamber. Then, the centrifugal force is increased to surpass the surface tension and the viscosity of the solution or melt; and the polymer jets exit from the nozzle tip, stretch, and finally the nanofibers deposit on the collector after solvent evaporation (Fig. 1.2e). The processing parameters including the solution concentration and viscosity, solvent evaporation rate, rotational speed, temperature, nozzle configuration, diameter, and length affect the fiber diameter and morphology [29]. Contrary to electrospinning, centrifugal spinning has a high capacity for scaling up at a low cost without the need for a high-voltage supply [48]. Besides, the fabrication of nanofiber from polymers can be performed at much higher concentrations in this method, resulting in lower solvent usage and lower production cost [49]. On the other hand, the spinneret design as well as polymer properties could affect the fiber quality and productivity in this technique [27].

Ren et al. [50] fabricated poly(L-lactic acid) (PLLA) and polyvinylpyrrolidone (PVP) composite fibers by centrifugal jet spinning process. Scanning electron microscopy (SEM) results showed that the two polymers were co-contiguously phase separated within the composite fibers during spinning and nanoscale roughness features were formed. They found that increasing PLLA surface roughness and porosity led to higher cell attachment and proliferation, which is required for tissue regeneration. Recently, Kwak et al. [51] designed an advanced centrifugal multispinning with a high production rate. They found that the contact angle and electrostatic charge of the multicomponent nanofiber could be justified by utilizing the co-spinning ability.

1.3.6 Blown Bubble-Spinning

Blown bubble-spinning as a low-cost simple process utilizes polymer melts or solutions to fabricate discontinuous nanofibers by using blowing air. The driving force in this method is based on a compressed gas (such as air or N_2) flowing through a concentric nozzle system [29]. In fact, the polymer solution/melt is injected through the inner nozzle at a constant injection rate, which is then extended by the high-pressure flow of a compressed gas blowing through the outer nozzle. By increasing the air pressure, the aerodynamic forces surpass the surface tension, and as a result, a jet is accelerated toward the collection target followed by solvent evaporation which results in the formation of micro- and nanofibers [48]. Figure 1.2f represents the schematic of blown bubble spinning. The most important factors in this procedure are the polymer concentration and molecular weight, gas pressure, injection rate, nozzle-collector distance, and nozzle configuration and diameter [52]. This technique is usually employed for materials with poor electrospinning ability having a low dielectric constant and/or electrical conductivity as well as polymers with high molecular weight (or high viscosity) that cannot be used in conventional electrospinning due to the high surface tension of their Taylor cone [29].

Tan et al. [53] developed composite multilayered filter masks from cellulose diacetate, poly(acrylonitrile) (PAN), and poly(vinylidene fluoride) by solution blow spinning (SBS) with high efficacy. In another study, Vasireddi et al. [54] designed a novel microfluidic gas flow-focusing nozzle for the fabrication of polymer nanocomposites. They claimed that the employed method resulted in a high production rate and direct fiber deposition without clogging. Besides, fiber morphology and surface patterning could be tailored by controlling the air pressure and polymer concentration. Recently, Sow et al. [55]

successfully recycled waste polystyrene to submicron fibers by using SBS. The enhanced hydrophobic and superoleophilic behavior of the produced fibers indicated their potential applications in oil-water separation. Even though centrifugal spinning and blown bubble-spinning is capable of obtaining high fiber productivity (high fiber production rate with low cost and easy fabrication), the production of nanofibers with diverse structures, i.e., core-shell or patterned structures, is still challenging. Besides, toxic solvent evaporation in this process is a major environmental concern [48].

1.3.7 Electrospinning

In electrospinning, which is one of the most flexible methods, electrical force is used for fabricating nanofibers. Basically, an electrospinning apparatus has three main elements: a) a high-voltage power supply, b) a spinneret/nozzle, and c) a grounded collector [39]. In this technique, a high voltage is employed between the nozzle and the collector. By increasing voltage up to a critical point, the charged fluid from the nozzle causes the formation of the Taylor cone, from which a jet is extended and nanofibers are fabricated and collected (Fig. 1.2g) [42]. Various kinds of collectors have been developed to tailor fiber orientation. Unaligned randomly oriented fibers are fabricated onto a static collector, whereas partially aligned nanofibers are produced by using a rotating collector [39, 56].

Two main factors influence the morphology and size of the electrospun nanofibers. The first one is the inherent characteristics of the polymer fluid its concentration, molecular weight, surface tension, conductivity, and rheological behavior. Another factor is operational conditions including the applied voltage, nozzle size, flow rate, nozzle-to-collector distance, and collector speed [57]. Electrospinning has several merits: (i) diverse types of materials can be used (e.g., polymers, carbon, ceramics, metal oxide, composite); (ii) commercial availability (easy and cheap) of setups for industrial production; (iii) feasible fabrication of different fibrous nanostructures with required dimension in one step process; (iv) functionalization of nanofibers could be performed either before or after spinning; and (v) nanofibers could be accumulated onto different materials (metal, fibrous mat, and glass) [29].

There is a wide range of electrospinning techniques including solution/melt electrospinning, multiple jet electrospinning, needless electrospinning, near-field electrospinning, bubble electrospinning, coaxial electrospinning, charge injection electrospinning, magnetic-assisted electrospinning, and nanospider electrospinning [58–61]. Core-shell nanofibers can be fabricated by using coaxial and triaxial electrospinning (Fig. 1.2h) [62]. For instance, Xu et al. [63] employed coaxial electrospinning for the fabrication of core-shell nanofibers composed of mesoporous silica nanoparticles as the core to encapsulate growth factors, and antibiotics-loaded nanofibers as the shell. According to the obtained results, the drug-loaded nanocomposite showed high potential for tissue regeneration.

Despite the extensive availability and low costs of electrospinning techniques for the fabrication of nanofibers, there is a limitation in mass production. Therefore, new solvent-free electrospinning techniques including hot melt electrospinning, UV-curing electrospinning, thermo-curing electrospinning, and supercritical CO_2-assisted electrospinning have been introduced [29]. However, specific and costly setups are required for these eco-friendly electrospinning methods. Moreover, larger nanofibers are produced using these techniques in comparison with the solution electrospinning process [59]. Huling et al. [64] succeeded in electrospinning of poly(ethylene glycol) diacrylate (PEGDA) with PVA under UV light for photo-cross-linking of the PEGDA. This reactive electrospinning is based on *in situ* cross-linking instead of solvent evaporation to stabilize fibers, which could decrease solvent usage. Recently, Zhao et al. [65] developed a portable melt electrospinning gun with no need for an external power supply. They employed a unique high heat transfer insulation part to resolve the problem of electrostatic interference. The apparatus is successfully used for electrospinning of polycaprolactone (PCL) onto wounds directly as a wound dressing.

1.3.8 Electro-centrifugal Spinning

Combining non-electrospinning techniques with electrospinning is one of the most recent attempts to increase the production rate of nanofibers, which has been made with the development of an electromechanical system. A new technique called electro-centrifugal spinning has emerged by coupling electrospinning with centrifugal spinning to fabricate nanoscale fibers [66, 67]. The setup includes a high-voltage supply with electrodes attached to the orifice(s) fixed on a rotating solution chamber and the inner wall of a cylinder centered on the solution chamber. Technically, by transferring a droplet of the polymeric solution to the tip of the orifice under centrifuge force, it is electrically charged immediately, and Taylor cone forms. Then both the electrical and the centrifugal field applied forces on the charged polymeric jet to stretch toward the interior surface of the cylinder as a collector. This unique combination of forces would result in a high alignment of the produced fibers as well as a higher rate of production at a lower applied voltage or slower rotating speed [67]. Basically, during the tension process, the jet passes a straight path toward the collector followed by a spiral path due to bending instability. Through this way, the jets are elongated by the repulsive force of charges on the surface. Fibers with micro- and nanoscale diameters are fabricated during the drying of jets to solidify and collect on the cylindrical collector (Fig. 1.2i) [68].

The ejected liquid jet is exposed to airflow as one of the main issues of electro-centrifugal spinning. The stream of rotating air surrounding the orifice induces rapid solvent evaporation in the jet and as a result, by causing difficulty in the jet elongation, thicker fibers are fabricated [69]. Dabirian et al. [66] developed a special electro-centrifugal spinning apparatus. To overcome the issue of surrounding airflow, they used a cylinder surrounding the nozzle and container to decrease the solvent vaporization rate. In another study, an air-sealed centrifuge electrospinning (ASCES) setup was prepared to overcome the air stream effect. At this apparatus, a mobile transparent door was used to inhibit airflow [67].

Different electro-centrifugal-spun fibers were produced from a different concentration of PAN solution (10–16 wt%) at a voltage of 15 kV and rotational speed of 6360 rpm. The results indicated that increasing solution concentration caused an increase in nanofibers' diameter [70].

Liu et al. [69] worked on the effect of collecting distance on polystyrene (PS) fibers alignment. They concluded that highly aligned straight fibers (97.7% of fibers within $\theta = 5°$) were fabricated from 18 wt% concentration of PS in tetrahydrofuran (THF) at the collecting distance of 2.5 cm. In another study, Khamforoush et al. [71] reported electro-centrifugal spinning of 17% PAN at 10–22 kV applied voltage, 1000–2000 rpm rotating speed, and 14 cm distance. According to the obtained results, the fibers' diameter decreased by increasing the rotational speed. Hosseinian et al. [72] assessed the impact of applied voltage (0–14 kV) as well as rotational speed (197–4051 r/min) on the morphology of PVP nanofibers. The results showed that under lower applied voltage and rotating speed only microparticles instead of fibers were produced, while by increasing the applied voltage, the minimum rotational speed required for the formation of a continuous jet declined.

1.4 Different Types of Nanofibrous Structures

There are various types of nanofibers based on their microscopic shape and morphology which can be confined by their degree of orientation, the amount and geometry of porosity, types, and several immiscible components, the presence of hollow channel(s) along the fibers, and so on [48]. Here, the most favorite types of nanofibers will be discussed.

1.4.1 Randomly Oriented Nanofibers

Basically, most conventional methods of nanofibrous fabrication (i.e., electrospinning, self-assembly, phase separation, etc.) lead to collecting a random morphology of nanofibers which are formed in random directions [73]. This kind of morphology is desired for lots of applications such as filtration, skin, and bone tissue engineering, separators, batteries, protective clothing, and so on. Randomly oriented nanofibrous mats are often created by the conventional electrospinning method which is widely employed to produce nano- to microscale fibers on a fixed flat-plate or low-speed drum as a collector. This fact is due to the bending instability of the jet through an electric field in the electrospinning process which causes the non-oriented deposition of dried nanofibers on the collector surface [74]. Figure 1.3a shows cardamom-loaded alginate-PVA nanofibers with random orientation for wound-healing application [15].

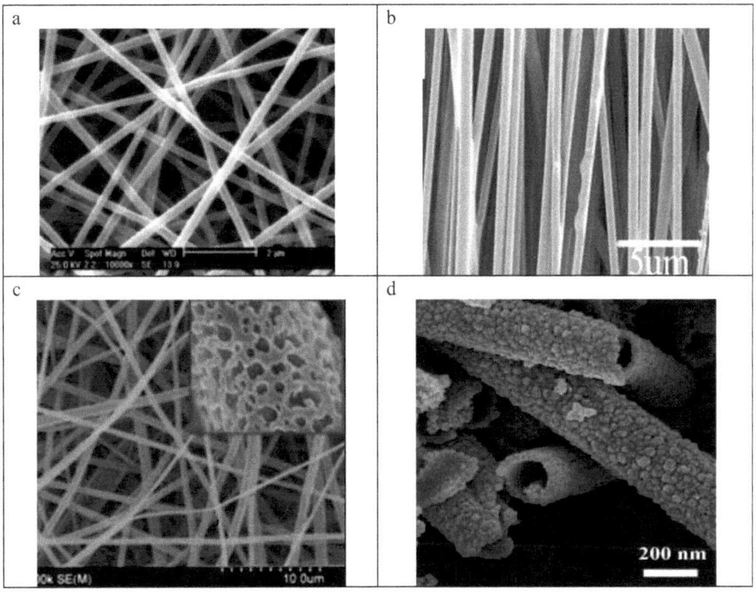

Figure 1.3 Different types of nanofibers: (a) random-oriented [15], (b) highly aligned nanofibers [48], (c) porous nanofibers [48], and (d) hollow nanofibers [48].

1.4.2 Aligned Nanofibers

Aligned nanofibers (Fig. 1.3b) have attracted lots of attention, especially in energy devices, (bio) sensors, and nerve tissue engineering. In the case of biomedical applications for neural tissue engineering, highly aligned nanofibrous scaffolds exhibit enhanced cell guiding, high migration, and differentiation as well as extracellular matrix (ECM) deposition because of the cellular tendency to be aligned along the nanofiber's direction. It also provides a further enhancement in mechanical properties compared to random nanofibrous webs [73].

Different modifications have been made to electrospinning to form aligned nanofibers which can be classified into three main groups according to the types of forces involved. The main involved forces in electrospinning modifications are mechanical, electrostatic, and magnetic forces [75]. It should be noted that these changes are mostly done on the collector system.

A typical form of involving mechanical force is using a high-speed rotating mandrel as a collector through which oriented nanofibers are formed under mechanical stretching along the rotating direction [76]. The alignment of electrospun nanofibers can also be enhanced by applying external electric or magnetic fields. In these categories, framework collectors are used in which there is a void gap between the conductive substrates and charged jets would be stretched across the gap and collected as dried nanofibers. The main framework collectors are rectangular framework, vertical parallel-plate, parallel-plate, and ring collectors. Besides, using electrode(s) coupling with drums as collectors called electrode-assistant collectors can also enhance the degree of fiber alignment in the electrospinning method [75, 76]. There are different electrode-assistant collector setups such as a single electrode, three-electrode, multi-electrode, and blade auxiliary electrode collectors.

1.4.3 Porous Nanofibers

Generally, porous materials are solid materials with porosity—the percentage of pores volume to the total volume—of 20% to 95%. According to the classification by IUPAC, porous materials fall into three main groups based on their average pore size: microporous (<2 nm), mesoporous (2 to 50 nm), and macroporous (>50 nm). Recently, nanoporous materials are defined as solids with a pore size of 0.2 to 50 nm which includes both micro and mesoporous [77]. Due to their more accessible active sites, numerous inner spaces followed by heterogeneous interfaces, and nanoporous materials have been considered for multifunctional applications such as filtrations, sensors, energy storage, drug delivery, and tissue engineering.

As it is known, random nanofibrous mats are considered porous materials in which pores or voids are created among the nanofibers. On the other hand, porous fibers are formed when the pores are created on individual fibers which offer interfiber pores (Fig. 1.3c). Porous nanofibers have attracted a lot of attention owing to their highest possible specific surface area and abundant pores which leads to excessive adsorption capacity. So, this unique combination of nanoporous with ultrafine fibers has introduced a promising candidate for catalysts, filtrations, supercapacitors, tissue engineering and drug delivery, energy storage systems, and so on [78].

There are four different mechanisms in porous nanofibers fabrications through electrospinning. The first mechanism is the *selective dissolution technique* in which the pores are generated along the electrospun nanofibers using an after-treatment process. In general, the after-treatment process is responsible for the removal of the excess polymer component or additives (salts, nanoparticles, etc.) from the main structure of nanofibers [77]. For example, Ji et al. [79] fabricated PAN porous nanofibers by removing silica nanoparticles from PAN-silica composite nanofibers using hydrofluoric acid (HF) as a solvent. The second mechanism is called *the selective pyrolysis composite formation* technique and its general concept is similar to the previous mechanism which is followed by thermal treatments such as carbonization, oxidation, and calcination as a post-treatment process. The most important product of this method is porous carbon nanofibrous web form PAN or PVP as precursors [80].

Breath figure and phase separation are the third and fourth mechanisms of porous nanofiber fabrication in which the generation of pores occurs during the electrospinning process in flight time of ejecting jet between nozzle and collector. The main concept of *the breath figure* is rapid solvent evaporation by cooling, in which moisture of air condenses on the surface of ejected jet and forms dews. During the drying of fiber, due to the convection current on the surface of the jet, the droplets of dew act as hard spheres which create pores on the surface of fibers [78]. Two main ways in *phase separation* mechanisms are TIPS and vapor-induced phase separation (VIPS). In TIPS, a rapid decrease in temperature followed by simultaneous solidification and sublimation of solvent causes thermodynamic instability which forms inter-porous electrospun nanofibers. The VIPS is based on penetration of non-solvent vapor—water vapor in most cases—and condensing in a liquid form through a wet electrospinning jet which causes liquid–liquid phase separation and ultimately after drying the jet pores will form along the nanofibers [77].

1.4.4 Hollow Nanofibers

Hollow nanofibers (Fig. 1.3d) are desired nanostructures for drug delivery and sensors, especially in gas sensing [81]. Electrospinning, template synthesis, and self-assembly are the three common methods for the fabrication of hollow nanofibers and the first two methods are much more used due to the ability to control the inner diameter of hollow fibers. In both template synthesis and electrospinning, the main step is preparing a continuous core-shell structure followed by post-treatment (i.e., thermal treatment or solvent extraction) for core removal. In template synthesis, the core or template usually is electrospun nanofibers and the shell is commonly formed by coating the electrospun nanofibers with other solutions. There are also various approaches in the electrospinning method to form hollow nanofibers including single spinneret, emulsion electrospinning, biaxial and three-axial electrospinning. Although the template synthesis method has been widely studied in different efforts, electrospinning is more popular owing to its simplicity and capability to form continuous hollow fibers [81].

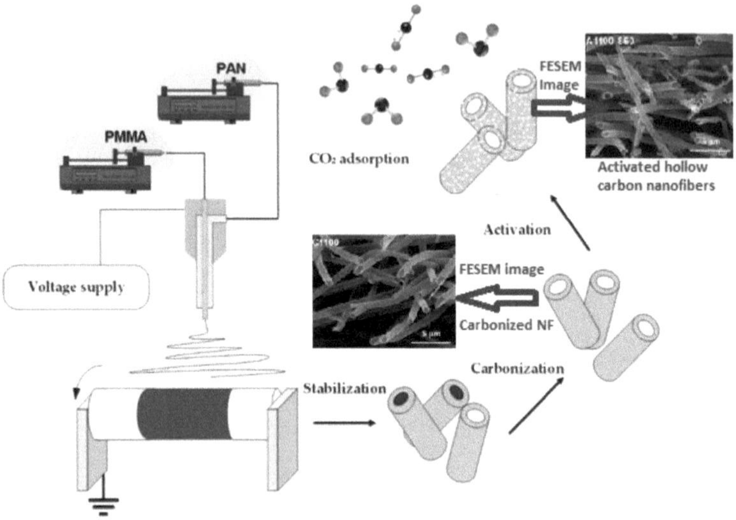

Figure 1.4 Fabricating process of hollow carbon nanofibers based on core-shell nanofibers from PMMA (core) and PAN (shell) [82].

Chiang et al. [82] fabricated hollow activated carbon nanofibers (HACNFs) for CO_2 capture, using coaxial electrospinning. PAN was employed as the shell and PMMA was used as the sacrificing core which was removed after stabilization during the carbonization stage. Figure 1.4 represents the processing stages in fabricating hollow activated carbon nanofibers.

1.5 Nanofibers for Technical Textiles

According to the unique properties of nanoscaled materials, coupling different properties of various polymers with nanofibrous structures leads to achieving high-tech materials desired for different purposes. Technical textiles which are based on or modified by nanofibrous layer(s) show considerable efficiency in different applications such as filtration, automotive textiles, geotextiles, and medical textiles.

1.5.1 Filtration

Filtration is one of the most popular applications of nanofibers for both gas and liquid filtration. Air and water pollution are the biggest environmental problems that are increasingly threatening human health [83]. Nowadays, respiratory infections along with gastrointestinal infections have great impacts on different common diseases and cancers which can be more controlled by removing harmful pollutants including fine particulate matter ($PM_{2.5}$[1]) and nitrogen dioxide (NO_2) from the air as well as nitrate, nitrite, and heavy metals like lead (Pb) from water [84, 85].

In the case of air filtration, many industries have used different approaches for cleaning the air and filtering out fine particles. For example, industries in biotechnology have adopted clean rooms to control contaminations. Moreover, humans spend a considerable part of their time indoors, so indoor air cleaning is a crucial activity and air filters have an important role in removing fine particles and toxic volatile organic compounds [83]. Various materials like foams and different types of fibers such as carbon fibers, fiberglass, cellulose, and spun-bond fibers individually or in combination with

[1]Very fine particles with a diameter of less than 2.5 μm which directly affect lung breathing

different kinds of additives have been investigated as air filters. Although fiber-based filters including glass fibers and melt-blown fibers have attracted more consideration due to easy processing and porous network structures, their large pore size and microscale fiber diameter have limited their applications especially in capturing aerosol and fine particles [86]. Regards, nanofiber-based filters are promising candidates owing to their high specific surface area together with nanoscale pore size. Many different nanofibers have been electrospun for filtration from natural polymers including proteins and polysaccharides, and synthetic polymers including polyethylene terephthalate (PET), polyacrylonitrile (PAN), polyvinyl alcohol (PVA), poly(lactic acid) (PLA), poly(ethylene glycol) (PEG), acrylonitrile butadiene styrene (ABS), polyurethane (PU), polyamide-6 (PA-6), polypropylene (PP), polystyrene (PS), etc. with or without additives like ZnO or silver nanoparticles as antibacterial agents and carbon nanotubes (CNTs) as absorbent [83]. Absorbent additives are added to the nanofiber substrate to enhance the adsorption capacity of contaminants. To achieve better physical-mechanical characteristics and higher efficiency in the separation of pollutants, a different mixture of polymers and additives as well as a number of modified electrospinning techniques have been studied in fabricating filters.

There are three main mechanisms for filtration according to medium filtration including capturing, deposition, and absorption. Moreover, the air filtration mechanism can be classified based on the pollutant types (i.e., particulate and gaseous pollutants) [83]. Particulate matters (PMs) can be trapped through fibrous filtration by inertial impaction, interception, diffusion, and electrostatic attraction (Fig. 1.5). Due to the higher amount of inertia of relatively large particles, they cannot control their direction to the fast deforming of flow lines. The large particles encounter directly with fibers, so the inertia impaction is the main filtration mechanism in this case. For finer particles that can easily adjust with streamlined movement, when the particles are close enough to the fiber surface, the interception mechanism is the dominant filtration process. The diffusion mechanism occurs due to Brownian motion based on the kinetic energy of particles when the fine particulate pollutants (with a diameter smaller than 100 nm) deviate from the streamlines. The electrostatic filtration mechanism is also performed when the attraction of asymmetric electrostatic charges between the particles and fibers occurs. The main effective parameters of nanofibrous filters on the filtration of particles are fiber diameters and distribution, particle and pore size, and airflow which are physical parameters. Besides, gaseous pollutants can be filtered based on the chemical properties of filter materials and the interaction between pollutants and filters.

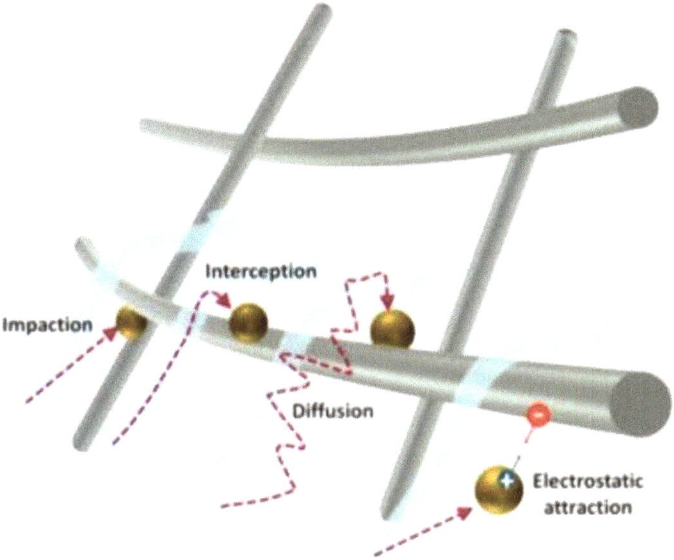

Figure 1.5 Schematic of capturing the process of PMs using nanofibrous filters [83].

Liu et al. [19] fabricated reusable bilayer nanofibrous filters from electrospun poly(methyl methacrylate)/poly(dimethylsiloxane) nanofibers as a hydrophobic layer and chitosan nanofibers as an antibacterial hydrophilic layer. They revealed that superhydrophilic chitosan fibers enhance the PM capture efficiency by up to 98% as well as antibacterial properties by about 95% against both gram-negative and gram-positive bacteria. The superhydrophobic PDMS/PMMA nanofibers in this filter play the role of a barrier as a moisture resistance to inhibit water stocking inside the membrane.

Li et al. [87] compared the type of polymers in nanofibrous filters on increasing the adsorption capacity of PMs. They revealed that the filtration efficiency of nanofibrous filters based on polycarbonate (PC) is considerably higher than those based on polyvinyl alcohol (PVA) and polystyrene (PS) which showed that higher polarity of PC nanofibrous filters lead to stronger inertial impaction mechanism followed by diffusion for better filtration.

In the case of water filtering, various techniques including chemical disinfectant treatments, distillation, sand filtration, membrane filtration, and reverse osmosis (RO) have been used, among which membrane filtration is the newest one. Therefore, among different types of membranes for water purification, RO would provide higher purification [88]. However, nanofiltration based on electrospun nanofibrous membranes has attracted more interest due to possessing a cost-effective, lightweight, and lower energy consumption procedure in comparison with conventional membranes. The approximate porosity of electrospun nanofibrous filters is about 80% which is effectively higher than the 5–35% porosity of conventional membranes. Although electrospun nanofibrous filters are considerably effective in purifying wastewater, severe fouling can be created during the filtration process due to their small pore size and distribution. Regards, surface modifications are highly considered to control the rate of fouling in nanofibrous filters. Electro-catalytic reduction is widely used for nitrate (NO_3^-) removal from water using different electrodes such as gold electrodes modified by polyaniline (PANI) nanofibers [89].

1.5.2 Automotive Applications

Reducing the weight and space occupied by the materials followed by increasing their mechanical strength is as of the main goals of various industries such as the automotive industry. Nanofibers and composites are the best options to improve conventional materials for the automotive textile industry. Textiles based on nanofibers possess small pore size, large surface area, and improved physical and chemical properties with fine fiber diameter. Textiles can be used in different parts of cars for various purposes such as composites for body parts, air filters, fuel filters, battery parts, and acoustic applications [90]. Nanofibers Generally, carbon nanofibers (CNFs), polymer-based nanofibers, and inorganic nanofibers have attracted more attention in automotive industries. Composites of CNFs with metal and polymer matrix provide acceptable mechanical properties with high thermal and electrical conductivity, which can be used in the outer panel of the car and bumpers to reduce weight and increase strength [91]. For instance, using 4 wt% of CNFs in the phenolic resin provides a high tensile strength of 43.8 MPa [91]. Polymeric nanofibers are mainly used in different car accessories such as cabin roofs, carpets, airbags, and filtration. Due to the porous structure of nanofibrous mats, they would act as noise reduction agents for acoustic application. As the porosity of nanofibrous mats is generally more than 90%, the pores act as traps, and by streaking the sound on the fibers, the kinetic energy of sounds will convert to heat and the sound will disappear. While conventional fibers like fiberglass or wool show less efficiency, especially in the low and moderate frequency range of sounds, nanofibrous layers can absorb 1000–3200 Hz sound with minimum thickness [92]. PVA or PAN nanofibers have been used on the carded web or spacer fabric for acoustic purposes [92, 93]. Ulrich et al. [93] fabricated nanofibrous membranes from polyamide 6 (PA6) and polyvinyl alcohol (PVA) on different porous acoustical materials such as polyester textiles using electrospinning. The results in both an impedance tube and a reverberation room showed significant improvement in noise reduction after incorporating the nanofibrous layer. Figure 1.6 shows the sound absorption coefficient as a function of frequency for polyester (PES) nonwoven fibrous bulk absorbers in different thicknesses (8 mm, 18 mm, and 46 mm) and the 8 mm one treated with 0.2 gsm PA6 nanofibrous layers. The result shows a great effect of an ultrathin PA6 nanofibrous layer on reducing noise.

Microcrystalline cellulose composites which are naturally obtained from the fibrous material of cellulose have high strength and modulus with low density and good thermal stability and can be combined with other synthetic polymers like polyamide 6 and polyesters. Due to their proper thermal stability, these nanocomposites can be used in body parts and the engine compartment of the car. On the other hand, tires also generate a high amount of heat due to their contact and friction with the road, so these kinds of nanofibrous composites can be promising candidates as main components of tires [94].

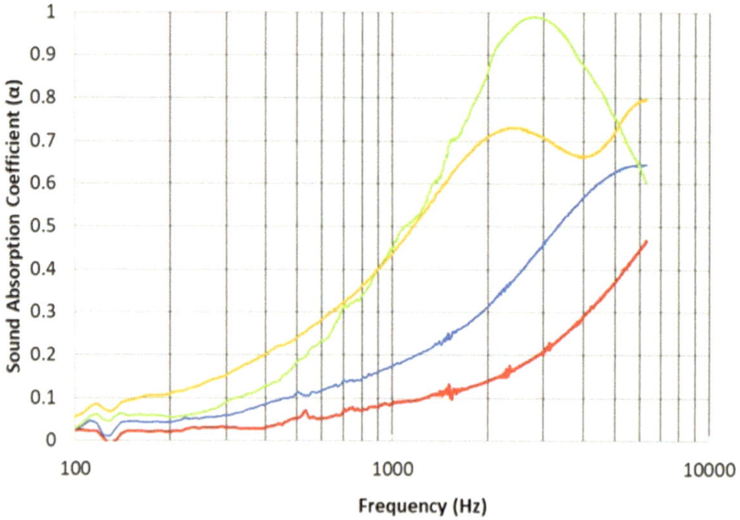

Figure 1.6 Normal incidence sound absorption coefficient for the PES nonwoven fibrous bulk absorber in different thicknesses: 8 mm (red curve), 18 mm (blue line), 46 mm (yellow line), and the 8 mm one treated with the 0.2 gsm PA6 nanofibrous membranes on the substrate (green line) [93].

One of the most applicable uses of nanofibers is in car air filters. The air filter with the intake provides appropriate cleanliness air supplied to the engine cylinders. Nevertheless, this leads to a continuous pressure drop on the filter which badly affects the engine work [95]. The value of the permissible filter resistance (Δp_{fdop}) determines the air filter ending service life. The durability of the filter is a technical term depending on the separation efficiency, pressure drop (car mileage), and filtration capacity. Some works have been performed to improve the filtering process of automotive air filters [96]. Dziubak et al. [96] reported a comparative study by adding a different fibrous layer to the standard cellulose paper filter. According to this study, the conventional paper filter was made of thicker cellulose fibers with an approximate diameter of 15 μm and in order to achieve higher filtration efficiency a nanofibrous layer was added to the base layer. The results showed that higher efficiency and filtration accuracy of dust grains below 5 μm was attained with the addition of nanofibers compared to standard filter paper.

1.5.3 Geotextiles

Geotextiles are woven or nonwoven fabrics based on polymeric fibers which are used in contact with soils and/or rock in the construction of roads, drains, harbor works, breakwaters, and other civil engineering purposes. Polypropylene (PP), PET, polyethylene (PE), and polyamides (PA) are the most common polymers used as geotextiles [97]. Although these polymers have considerable advantages for geotextile applications, they need modifications due to some drawbacks such as poor sensitivity of PP, easy hydrolysis of PET in the soil in basic media, lack of supply of PE fibers, and poor comprehensive performance of PA. To enhance the performance of geotextiles, some additives like antioxidants, light, and thermal stabilizers, UV absorbers, flame retardants, lubricants, and antibacterial agents are added to the polymeric substrate of geotextiles. However, using green geotextiles based on natural polymeric fibers such as cellulose, hemicellulose, lignin, and pectin is expanding. Since the properties of natural fibers vary with different types, fibers with higher mechanical properties such as jute and coir fibers are chosen as geotextile raw materials [98].

Recently, a considerable number of high-performance geotextiles such as wicking geotextiles [99] and basalt fiber needle mats [100] have been introduced for different functions. Generally, the main functions of geotextiles are separation, filtration, drainage, and reinforcement (Fig. 1.7) [97]. Geotextiles can be used to separate different materials in contact, to avoid mixing with each other and failing their structural integrity. For example, if no geotextile is used between stone aggregates and fine-grained soil in the lower layer, the soil tries to enter the voids among stones and will lose its drainage capacity. On the other hand, the aggregate stone of the upper layer will settle into the soft-grained soil and lose its strength. The filtration function of geotextiles is important to prevent soil migrations through the water flow during seepage and infiltrating drainage. So, it will support the soil structure from collapsing. Moreover, geotextiles are well-known for water discharging and play an important role as a drainage channel. Due to their high tensile modulus, tensile strength, and surface friction, geotextiles are also used in soil reinforcement to enhance the strength and deformation performance of reinforced composite soils [97].

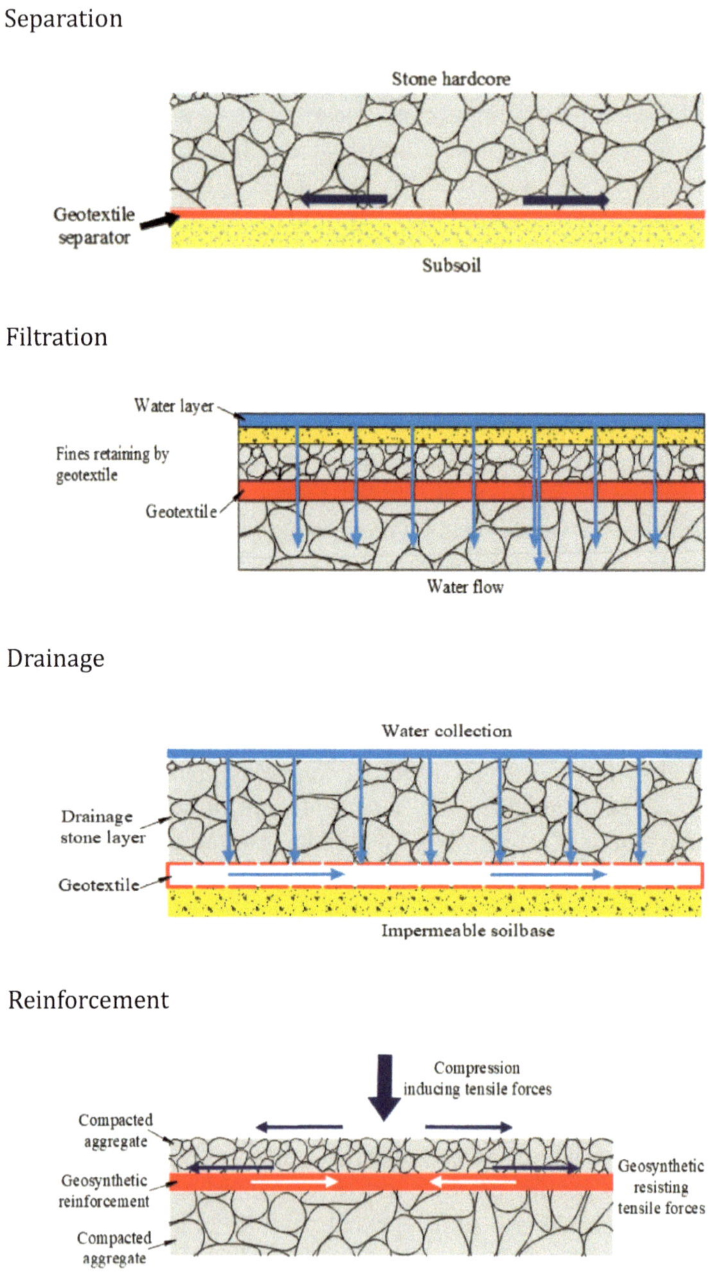

Figure 1.7 Schematic of main functions of geotextiles [97].

Generally, to improve the filtration function, higher porosity as well as higher specific surface area are needed. In fabricating geotextiles using conventional fibers, more fibers would be used per unit area to provide more adsorbent sites, but this method leads to reduce porosity. Regards, nanofibers due to their unique characteristics can improve the desired functions of geotextiles, remarkably. In separation and filtration functions, nanofibers can provide a smaller pore size as well as a higher specific surface area because more fibers can be integrated into a given space and improve filtration and separation efficiency [101].

1.5.4 Medical Textiles

The high surface area of nanofibers promotes cell attachments and proliferation required for biomedical applications. Different types of biopolymers have been electrospun into nanofibers. Natural polymers from renewable resources such as plants and animals are divided into two main groups: polysaccharides (cellulose, alginate, chitin/chitosan, hyaluronic acid, etc.) as well as proteins (collagen, gelatin, keratin, fibroin, etc.). Most natural polymers are highly biocompatible but suffer from low mechanical strength. Synthetic polymers such as poly(vinyl alcohol) (PVA), poly(ethylene glycol) (PEG), poly(lactide) (PLA), and poly(caprolactone) (PCL), etc. have usually higher mechanical strength than natural polymers, but their biological performance is limited. Therefore, they are combined with natural polymers or functionalized to improve their biocompatibility [102]. Nanofibrous medical textiles can be classified into three groups: wound dressing, tissue engineering scaffolds, as well as drug delivery, which will be explained in the following sections.

1.5.4.1 Wound dressing

Skin as the largest organ of the human body provides a protective barrier against microbial, chemical, mechanical, and thermal damage. An ideal scaffold for skin tissue engineering should possess biocompatibility, biodegradability, and proper mechanical properties. Wound dressings are used to assist rapid healing at a fair cost causing minimum inconvenience to the patients [103]. The healing process for acute wounds including surface burns, chemical injuries, mechanical injuries, and surgical wounds, follows the normal wound healing cycle within 8–12 weeks. While chronic wounds which occur mainly due to certain specific diseases (such as diabetes) carry on for more than 12 weeks and may reoccur. Chronic wounds are at risk of infection by inflammatory bacteria affecting wound repair [104]. Wound dressings are employed to protect the wound from the external environment, provide a moist environment with effective oxygen and water vapor permeability, absorb the excess of wound exudates, and enhance the healing process [105].

Nanofibers have excellent properties in promoting wound healing. Their large surface area, porous structure, and high permeability can absorb the exudate and provide a moist environment for wound healing. The morphology and dimension of nanofibers bear a resemblance to local ECM structure which can enhance cell adhesion, migration, and proliferation. In addition, different types of antimicrobial agents, growth factors, and therapeutic agents can be loaded into nanofibers to promote skin regeneration and inhibit contamination [29]. Nowadays, there is a wide range of functionalized bioactive wound dressings that can release antimicrobial agents to accelerate wound healing. The bioactive nanofiber dressings can provide the requirements of the healing process, promote cell growth, and prevent inflammation [104]. Plant extracts have received great attention in wound dressings due to their suitable biological properties. For instance, Najafi et al. [15] loaded cardamom extract into sodium alginate and PVA nanofibers to induce antibacterial activity. *In vitro* assays showed that the electrospun nanofibers had a high potential for antimicrobial wound dressings. Azarniya et al. [106] developed composite scaffolds composed of nanofibers (keratin/bacterial cellulose) and thermoresponsive hydrogel (Tragacanth gum/Pluronic) by concurrent electrospinning and electrospraying technique. The hydrogel particles were uniformly embedded into the electrospun nanofibers at the junctions without hampering their porosity. In another study, Zahedi et al. [107] fabricated core-shell electrospun nanofibers incorporated by aloe vera extract as an active scaffold for wound healing. *In vitro* assay showed an enhancement in cellular attachment and growth with aloe vera incorporation having no cytotoxic effects. Recently, Samadian et al. [108] successfully produced antibacterial cellulose acetate/gelatin nanofibers embedded with berberine for diabetic foot ulcer treatment. *In vitro* studies showed that the proliferation of the L929 fibroblastic cells was promoted in the presence of berberine. Besides, *in vivo* assays on diabetic rats confirmed that the fabricated nanofibers were good candidates for wound dressings.

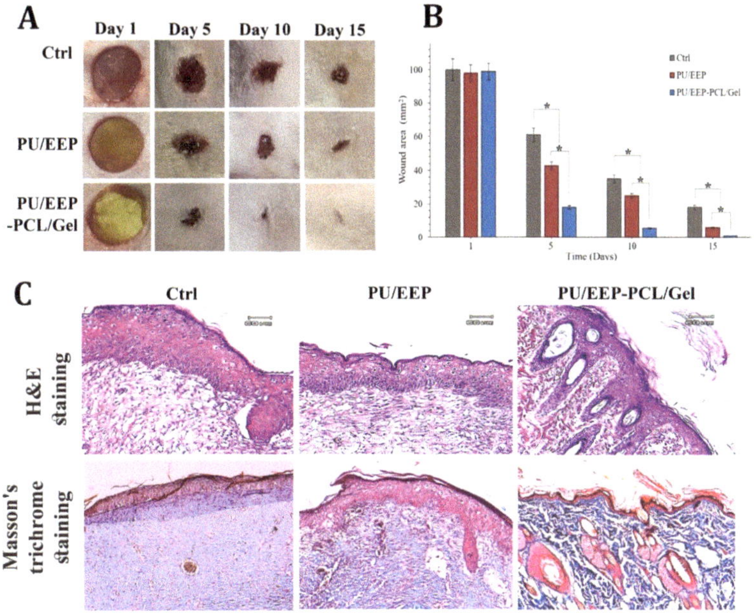

Figure 1.8 (A) Macroscopic photographs of the wounds at definite time points (1st, 5th, 10th, and 15th day after wound creation) to exhibit wound closure progression at different groups (Ctrl, PU/EEP, and PU/EEP-PCL/gel). (B) Wound closure progression according to the calculation of the remaining wound area at definite time points (1st, 5th, 10th, and 15th day after wound creation). The data are presented as mean ± SD (n = 8). (C) Photographs of the H&E and Masson's trichrome stained sections of the healed skin specimens on the 15th day after wound creation [110].

In situ electrospinning is a novel technique for fabricating wound dressing. This strategy is more convenient and can be adjusted easily according to the patient's needs [104]. Dong et al. [109] developed a new antibacterial dressing based on polyvinyl pyrrolidone containing is at its root via a handheld electrospinning apparatus directly on a wound site. Results showed wound closure after 11 days in a mouse skin injury mode and epidermal repair in histological studies. In another study, Eskanradinia et al. [110] fabricated a bilayer wound dressing made of polycaprolactone/gelatin (PCL/

gel) electrospun web and a dense membrane of polyurethane–ethanolic extract of propolis (PU/EEP). The top layer (PU/EEP) was used as an antibacterial agent and the sublayer (electrospun PCL/gel) was used to enhance cell attachments in the wound bed. The fabricated dressing showed high hydrophilicity, biocompatibility, and antibacterial activity to protect the wound from infection in the Wistar rat model (Fig. 1.8). As can be seen in Fig. 1.8, the *in vivo* study showed high potential of the bilayer dressing in wound healing due to its significant effect on wound closure and collagen deposition.

1.5.4.2 Tissue-engineered scaffolds

Tissue engineering (TE) requires cells, scaffolds, and biochemical and biophysical factors to regenerate functional tissues. Nanofibrous polymer composites have been employed as potential TE scaffolds owing to their high surface area, excellent mechanical properties, flexibility in surface functionalities, high porosity and permeability, and promoted cell adhesion and growth [111]. In recent years, a wide range of both natural and synthetic polymers and their blends have been used to fabricate nanofibers as TE scaffolds for various tissues [29]. In fact, the microstructure, shape, size, and geometrical distribution of the electrospun nanofibers resemble the collagen fibrils in the natural ECM, leading to improved cell attachment and proliferation [28]. The biomaterial composition, pore structure, mechanical properties, surface characteristics, and biodegradability of the polymeric nanofibers could be tailored to meet the requirements for specific tissue regeneration. Highly aligned nanofibers have usually better mechanical properties and biological performance in comparison with random nanofibers [48].

Cardiovascular disease is one of the main causes of mortality globally. Electrospinning has attracted attention in the development of tubular fibrous membranes as tissue-engineered vascular grafts [112]. The mechanical performance of electrospun blood vessels plays a key role in practical applications since they should resist frequent blood circulation and the corresponding pressure. Besides, to inhibit platelet activation, the anti-thrombogenic activity of the fabricated vascular grafts is essential [113]. Although PCL is the most commonly used polymer for this purpose, its hydrophobicity can be surpassed by combining it with natural polymers or bioactive materials. Pan et al. [114] mixed PCL with polydioxanone (PDS) to enhance the mechanical properties as well as the hydrophilicity of the composite in comparison with pristine PCL. In another study, Xing et al. [115] coated electrospun PCL vascular grafts with gelatin in order to promote endothelialization, which is vital for preventing intimal hyperplasia in the long term.

Bone is a vascular and highly specialized form of connective tissue consisting of organic (collagen nanofibers) and inorganic (nanocrystalline hydroxyapatite) constituents in a hierarchical structure. Although there have been great advances in bone TE, more efforts are still required for the development of novel biomimetic scaffolds to regenerate nanoscale topographical and biofactor cues in natural bone tissue [116]. Electrospun scaffolds are extensively employed for bone TE owing to their unique properties including high porosity with interconnected pores, osteoinductivity, biodegradability, and mechanical stability [28]. Nedjari et al. [117] reported the fabrication of a 3D honeycomb electrospun scaffold composed of poly (L-lactide-ε-caprolactone) and fibrinogen. The results showed the scaffolds significantly supported the osteogenic differentiation of human adipose-derived mesenchymal stem cells (hMSCs) for bone reconstruction. Similarly, Samadian et al. [118] developed a 3D scaffold consisting of poly (L-lactic acid)/PCL matrix polymer embedded with electrospun gelatin nanofibers and gold nanoparticles. *In vitro* as well as *in vivo* assays revealed that the produced scaffold had excellent biocompatibility, osteoconductivity, and mechanical properties required for bone TE.

Cartilage is a flexible tissue between the bones composed of collagen fibers, proteoglycans, and glycosaminoglycans. Cartilage defects therapy is one of the most challenging clinical problems in orthopedics because of its low repair capacity [119]. Development of biomimetic tissues capable of recapitulating the biological, structural, and functional properties of native cartilage, preventing the progress of cartilage atrophy, and enhancing long-term functional effects for patients have been attempted by many researchers [120, 121]. Nanocomposite scaffolds made of nanofibers embedded in the hydrogel matrix have been attempted to closely mimic the natural ECM [122]. For instance, Liu et al. [123] developed 3D scaffolds based on poly(lactic-co-glycolic acid) (PLGA) electrospun nanofiber embedded hydroxybutyl chitosan hydrogels, which were injected into the 3D printing PCL framework (Fig. 1.9). The produced scaffolds showed considerably enhanced chondrogenesis and mechanical properties along with adjustable internal microchannels for cartilage differentiation of hMSCs. Moradi et al. [23] also investigated the effect of bulking process (ultrasonication, followed by freeze-drying) on the electrospun PCL scaffolds incorporating graphene oxide (GO). The obtained results revealed bulking process led to a higher porosity of the nanofibers. *In vivo* assay also showed highly qualified healing of cartilage defect on a rabbit model. Recently, Sun et al. [124] developed pH-triggered biphenyl-tripeptide nanofiber supramolecular hydrogels which could support chondrocytes adhesion and growth, and promote chondrogenic matrix secretion.

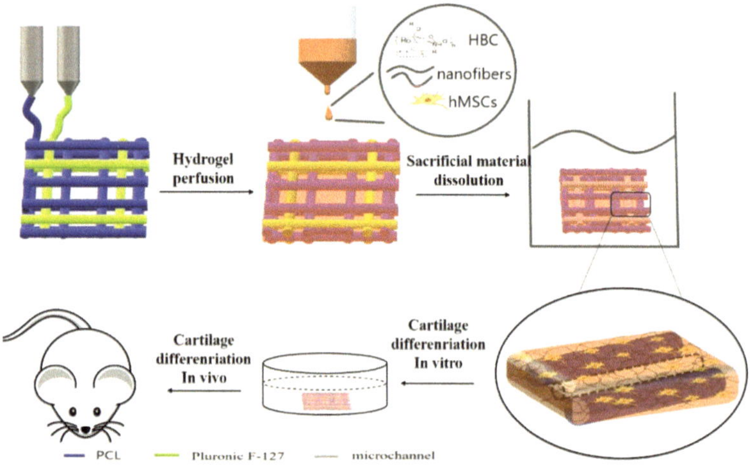

Figure 1.9 Schematic illustration of 3D PCL scaffold reinforced with PLGA electrospun nanofiber for cartilage regeneration [123].

The disruption of axonal bundles in neural injuries is a major issue in healthcare [125]. Yao et al. [126] designed 3D hierarchically aligned fibrin hydrogel by simultaneously electrospinning and self-assembly process. The scaffold was successfully employed in a rat dorsal hemisected spinal cord injury model to bridge the lesion site and accelerate axonal regrowth. Electroactive biomaterials can provide an electrical stimulus to regulate neuronal activity in neural TE. Steel et al. [127] fabricated hyaluronic acid-carbon nanotube nanofibers by electrospinning and found that electrical stimulation could enhance neurite outgrowth and activate voltage-sensitive regenerative pathways. Another approach is to culture neural stem cells (NSCs) on conductive substrates to induce differentiation. Garrudo et al. [128] developed electrically conductive PCL-polyaniline electrospun scaffolds for neural TE. The results showed that the fabricated nanofibers could support NSCs attachment and proliferation without any major changes in the cell's morphology.

1.5.4.3 Drug delivery

To overcome the problems of systemic administration, controlled drug delivery to a particular target in a sustained manner has been attempted. The capability of a drug delivery system (DDS) relies on the specifications of the drug vehicles, which should have biocompatibility, high drug-loading capacity, desirable drug-release profile, and effective cellular uptake [48]. Nanofibers have attracted great attention in DDS, owing to their easy fabrication, wide range of applicable materials, and drug encapsulation [29]. The high specific surface area and the easy modification of the employed polymers in nanofiber result in higher encapsulation efficiency and drug-loading capacity [129]. Various bioactive molecules and therapeutic agents can be embedded into the electrospun nanofibers or on their surfaces by post-treatment, co-electrospinning of drug-polymer materials, coaxial/emulsion electrospinning, etc. Among these different methods, co-electrospinning is the preferred strategy for drug loading owing to its relatively low cost and simplicity [130]. It should be noted that the drug-release profile on polymeric nanofibers could be tailored by selecting the spinning production technique, fiber structures, and material composition [29]. Among different fiber structures, core-shell nanofibers are the most suitable for controlled drug delivery, because the drug release can be tailored by choosing the inner and outer layers components. Moreover, the co-incorporation of multiple therapeutic agents with varied solubility properties can be performed in one step [48].

Recently, stimuli-responsive polymer-based nanofibers have received great attention in targeted DDS to release the drug in response to physical, biological, and/or chemical stimuli [129]. Nanoassemblies sensitive to external stimuli, e.g., light, magnetic field, pH, or temperature, have been attempted for remotely controlled drug delivery [131–133]. These systems provide a targeted release of the drug, leading to an adjustment of the required amount of the drug based on the patient's needs, enhancing therapeutic potency and decreasing the side effects [134]. In fact, the higher surface area/volume ratio of functional smart nanofibers results in the controlled and triggered release of therapeutic agents with higher efficacy. Amarjargal et al. [135] developed polymethacrylate-based thermoresponsive nanofibers for drug delivery. They found no model drug was released at temperatures below the glass temperature owing to the limited movement of the macromolecular chains. While by increasing temperature higher than glass temperature, the drug was released in a controlled manner with prolonging time (Fig. 1.10). Schoeller et al. [136] coated PLGA nanofibers with chitosan and alginate layers to induce pH-sensitivity for ibuprofen release. According to the obtained results, drug release was decreased in acidic pH in comparison to neutral pH owing to the possible interactions between the coating and the drug. In another study, Banerjee et al. [137] introduced a dual magneto- and opto-stimuli-responsive DDS composed of functionalized electrospun nanofibers incorporated with superparamagnetic iron oxide nanoparticles. They found that the structural changes of the thermosensitive nanofibers controlled the drug-release profile depending on the nanoparticles' loading and duration of stimulation.

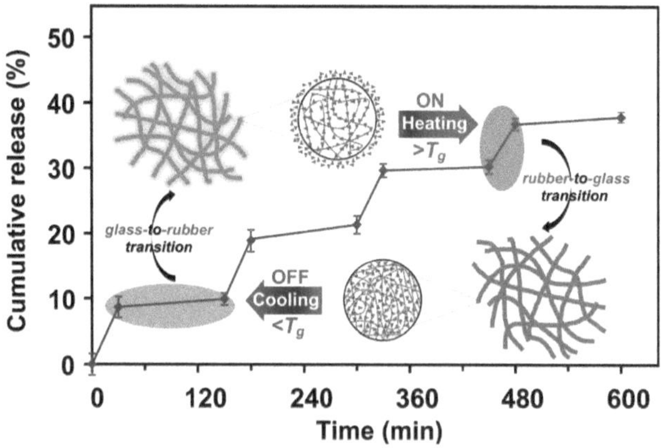

Figure 1.10 Drug-release profile for thermoresponsive electrospun nanofibers with on–off switching cap [135].

Although, there is significant progress in triggered nanofiber-based DDS, *in vitro* studies have been attempted in most research, and clinical translation of such technologies requires more development. The main limitation is ascribed to the biological responses, chemical intricacy, and predictability of stimulated drug-release carriers in the physiological environment *in vivo* [48].

1.6 Conclusion and Future Perspectives

Polymeric nanofibers have attracted more and more attention in recent years due to their large specific surface area, controllable fiber morphology, and high porosity. In this chapter, the progress of fabrication techniques to produce nanofibers from different types of natural and synthetic polymers in different structures has been reviewed to accomplish the requirements of a particular application. As for biomedical applications, the fabrication of an ideal 3D biomimetic scaffold for a specific tissue and the translation of these technologies into the clinic is still challenging. There is an urgent need to develop innovative spinning strategies to produce nanofibers using cost-effective manufacturing scale-up with mass production in the industry having better control of the features at the nanoscale. It is expected that conventional electrospinning methods will be merged with 3D printing or textile technologies such as weaving, braiding, and embroidering in near future.

References

1. Jeevanandam, J., et al. (2018). Review on nanoparticles and nanostructured materials: history, sources, toxicity and regulations, *Beilstein J Nanotechnol*, **9**, pp. 1050–1074.
2. Shi, H., et al. (2013). Titanium dioxide nanoparticles: a review of current toxicological data, *Part. Fibre Toxicol.*, **10**, pp. 15.
3. Pokropivny, V.V. and V.V. Skorokhod. (2007). Classification of nanostructures by dimensionality and concept of surface forms engineering in nanomaterial science, *Mater Sci Eng C*, **27**, pp. 990–993.
4. Tiwari, J.N., R.N. Tiwari, and K.S. Kim. (2012). Zero-dimensional, one-dimensional, two-dimensional and three-dimensional nanostructured

materials for advanced electrochemical energy devices, *Prog Mater Sci*, **57**, pp. 724–803.

5. Wang, X., et al. (2014). One-dimensional titanium dioxide nanomaterials: nanowires, nanorods, and nanobelts, *Chem Rev*, **114**, pp. 9346–9384.
6. Mann, A.K.P. and S.E. Skrabalak. (2011). Synthesis of single-crystalline nanoplates by spray pyrolysis: a metathesis route to Bi2WO6, *Chem Mater*, **23**, pp. 1017–1022.
7. Siril, P.F., et al. (2009). Synthesis of ultrathin hexagonal palladium nanosheets, *Chem Mater*, **21**, pp. 5170–5175.
8. Vizireanu, S., et al. (2010). Plasma techniques for nanostructured carbon materials synthesis. A case study: carbon nanowall growth by low pressure expanding RF plasma, *Plasma Sources Sci Technol*, **19**, pp. 034016.
9. Kwon, Y. and S. Kim. (2021). Indium phosphide magic-sized clusters: chemistry and applications, *NPG Asia Mater*, **13**, pp. 37.
10. Wang, L. and Y. Yamauchi. (2009). Facile Synthesis of three-dimensional dendritic platinum nanoelectrocatalyst, *Chem Mater*, **21**, pp. 3562–3569.
11. Abdul Sammed, K., et al. (2021). Carbon nanocoil-supported three-dimensional structure of nickel–cobalt nitrides as the electrode material for supercapacitors, *ACS Appl Energy Mater*, **4**, pp. 6678–6687.
12. Wang, H., et al. (2016). The morphologies and optical properties of three-dimensional GaN nano-cone arrays, *RSC Adv*, **6**, pp. 43272–43277.
13. Yilmaz, C., et al. (2014). Three-dimensional crystalline and homogeneous metallic nanostructures using directed assembly of nanoparticles, *ACS Nano*, **8**, pp. 4547–4558.
14. Yi, W., et al. (2022). Three-dimensional flower-like nickel oxide/graphene nanostructures for electrochemical detection of environmental nitrite, *ACS Appl Nano Mater*, **5**, pp. 216–226.
15. Najafi, S., et al. (2021). Study on release of cardamom extract as an antibacterial agent from electrospun scaffold based on sodium alginate, *J Text Inst*, **112**, pp. 1482–1490.
16. Zdraveva, E., et al. (2017). 11 - Electrospun nanofibers, in *Structure and Properties of High-Performance Fibers*, G. Bhat, ed., Woodhead Publishing: Oxford, p. 267–300.

17. Gholipour-Kanani, A.H.B. (2010). Review on electrospun nano fibrous scaffold and their biomedical applications, *Trends in Biomater Artif Organs*, **24**, pp. 93–115.
18. Xue, J., et al. (2019). Electrospinning and electrospun nanofibers: methods, materials, and applications, *Chem Rev*, **119**, pp. 5298–5415.
19. Liu, H., et al. (2019). Transparent antibacterial nanofiber air filters with highly efficient moisture resistance for sustainable particulate matter capture, *iScience*, **19**, pp. 214–223.
20. Jafari, G., et al. (2021). Novel non-enzymatic glucose biosensor based on electrospun PAN/PANI/CuO nano-composites, *J Text Inst*, pp. 1–8.
21. Manatunga, D.C., et al. (2020). Nanofibrous cosmetic face mask for transdermal delivery of nano gold: synthesis, characterization, release and zebra fish employed toxicity studies, *R Soc Open Sci*, **7**, pp. 201266.
22. Xu, H., et al. (2020). Phage nanofibers in nanomedicine: biopanning for early diagnosis, targeted therapy, and proteomics analysis, *Wiley Interdiscip Rev Nanomed Nanobiotechnol*, **12**, pp. e1623.
23. Moradi, F., et al. (2021). The effect of bulk electrospun polycaprolactone-graphene oxide scaffold on the healing of defected femur cartilage on a rabbit model, *Fibers Polym*, **22**, pp. 1247–1255.
24. Gholipour-Kanani, A., et al. (2018). Poly (ε-caprolactone)–chitosan-poly (vinyl alcohol) nanofibrous scaffolds for skin excisional and burn wounds in a canine model, *IET Nanobiotechnol*, **12**, pp. 619–625.
25. Kajdič, S., et al. (2019). Electrospun nanofibers for customized drug-delivery systems, *J Drug Deliv Sci Technol*, **51**, pp. 672–681.
26. Wang, Y., T. Yokota, and T. Someya. (2021). Electrospun nanofiber-based soft electronics, *NPG Asia Mater*, **13**, pp. 22.
27. Alghoraibi, I. and S. Alomari. (2018). Different methods for nanofiber design and fabrication, in *Handbook of Nanofibers*, Barhoum, A., M. Bechelany, and A.S.H. Makhlouf, eds., Springer Nature Switzerland AG, pp. 1–46.
28. Islam, M.S., et al. (2019). A review on fabrication of nanofibers via electrospinning and their applications, *SN Appl Sci*, **1**, pp. 1–16.
29. Barhoum, A., et al. (2019). Nanofibers as new-generation materials: from spinning and nano-spinning fabrication techniques to emerging applications, *Appl Mater Today*, **17**, pp. 1–35.
30. Zhao, J., et al. (2011). Preparation, structure and crystallinity of chitosan nano-fibers by a solid–liquid phase separation technique, *Carbohydr Polym*, **83**, pp. 1541–1546.

31. Liu, X. and P.X. Ma. (2009). Phase separation, pore structure, and properties of nanofibrous gelatin scaffolds, *Biomaterials*, **30**, pp. 4094–4103.
32. Li, X.-t., Y. Zhang, and G.-Q. Chen. (2008). Nanofibrous polyhydroxyalkanoate matrices as cell growth supporting materials, *Biomaterials*, **29**(27), pp. 3720–3728.
33. Liu, X. and P.X. Ma. (2010). The nanofibrous architecture of poly(L-lactic acid)-based functional copolymers, *Biomaterials*, **31**, pp. 259–269.
34. Chen, V.J. and P.X. Ma. (2004). Nano-fibrous poly(L-lactic acid) scaffolds with interconnected spherical macropores, *Biomaterials*, **25**, pp. 2065–2073.
35. Samitsu, S., et al. (2013). Flash freezing route to mesoporous polymer nanofibre networks, *Nat Commun*, **4**, pp. 2653.
36. Bajakova, J., et al. Drawing: the production of individual nanofibers by experimental method, in *Proceedings of the 3rd International Conference on Nanotechnology: Smart Materials* (NANOCON'11), 2011.
37. Ma, J., et al. (2016). A rapid and simple method to draw polyethylene nanofibers with enhanced thermal conductivity, *Appl Phys Lett*, **109**, pp. 033101.
38. Yadavalli, N.S., et al. (2020). Gravity drawing of micro- and nanofibers for additive manufacturing of well-organized 3D-nanostructured scaffolds, *Small*, **16**, pp. 1907422.
39. Gugulothu, D., et al. (2019), Fabrication of nanofibers: electrospinning and non-electrospinning techniques, in *Handbook of Nanofibers*, Barhoum, A., M. Bechelany, and A.S.H. Makhlouf, eds., Springer International Publishing: Cham, p. 45–77.
40. Tao, S.L. and T.A. Desai. (2007). Aligned arrays of biodegradable poly(ε-caprolactone) nanowires and nanofibers by template synthesis, *Nano Lett*, **7**, pp. 1463–1468.
41. Liao, H.-S., et al. (2016). Self-assembly mechanisms of nanofibers from peptide amphiphiles in solution and on substrate surfaces, *Nanoscale*, **8**, pp. 14814–14820.
42. Zhang, X. and Y. Lu. (2014). Centrifugal spinning: an alternative approach to fabricate nanofibers at high speed and low cost, *Polym Rev*, **54**, pp. 677–701.
43. Beniash, E., et al. (2005). Self-assembling peptide amphiphile nanofiber matrices for cell entrapment, *Acta Biomater*, **1**, pp. 387–397.

44. Hartgerink, J.D., E. Beniash, and S.I. Stupp. (2001). Self-assembly and mineralization of peptide-amphiphile nanofibers, *Science*, **294**, pp. 1684–1688.

45. Chen, W., et al. (2019). Self-assembled peptide nanofibers display natural antimicrobial peptides to selectively kill bacteria without compromising cytocompatibility, *ACS Appl Mater Interfaces*, **11**, pp. 28681–28689.

46. Ma, X., et al. (2021). Chitosan/polydopamine layer by layer self-assembled silk fibroin nanofibers for biomedical applications, *Carbohydr Polym*, **251**, pp. 117058.

47. Cui, T., et al. (2020). Instant self-assembly peptide hydrogel encapsulation with fibrous alginate by microfluidics for infected wound healing, *ACS Biomater Sci Eng*, **6**, pp. 5001–5011.

48. Dos Santos, D.M., et al. (2020). Advances in functional polymer nanofibers: from spinning fabrication techniques to recent biomedical applications, *ACS Appl Mater Interfaces*, **12**, pp. 45673–45701.

49. Zhao, Y., et al. (2016). Preparation of nanofibers with renewable polymers and their application in wound dressing, *Int J Polym Sci*, **2016**, pp. 4672839.

50. Ren, L., et al. (2013). Large-scale and highly efficient synthesis of micro- and nano-fibers with controlled fiber morphology by centrifugal jet spinning for tissue regeneration, *Nanoscale*, **5**, pp. 2337–2345.

51. Kwak, B.E., et al. (2021). Large-scale centrifugal multispinning production of polymer micro- and nanofibers for mask filter application with a potential of cospinning mixed multicomponent fibers, *ACS Macro Lett*, **10**, pp. 382–388.

52. Medeiros, E.S., et al. (2009). Solution blow spinning: a new method to produce micro- and nanofibers from polymer solutions, *J Appl Polym Sci*, **113**, pp. 2322–2330.

53. Tan, N.P.B., et al. (2019). Solution blow spinning (SBS) nanofibers for composite air filter masks, *ACS Appl Nano Mater*, **2**, pp. 2475–2483.

54. Vasireddi, R., et al. (2019). Solution blow spinning of polymer/nanocomposite micro-/nanofibers with tunable diameters and morphologies using a gas dynamic virtual nozzle, *Sci Rep*, **9**, pp. 14297.

55. Sow, P.K., Ishita, and R. Singhal. (2020). Sustainable approach to recycle waste polystyrene to high-value submicron fibers using solution blow spinning and application towards oil-water separation, *J Environ Chem Eng*, **8**, pp. 102786.

56. Sahay, R., V. Thavasi, and S. Ramakrishna. (2011). Design modifications in electrospinning setup for advanced applications, *J Nanomater*, **2011**, pp. 317673.
57. Li, D. and Y. Xia. (2004). Electrospinning of nanofibers: reinventing the wheel?, *Adv Mater*, **16**, pp. 1151–1170.
58. Bubakir, M.M., et al. (2019), Advances in melt electrospinning technique, in *Handbook of Nanofibers*, Barhoum A., M. Bechelany, and A.S.H. Makhlouf, eds., Springer International Publishing: Cham, p. 125–156.
59. Zhang, B., et al. (2017). Solvent-free electrospinning: opportunities and challenges, *Polym Chem*, **8**, pp. 333–352.
60. Moon, S., G. Manjae, and K. Lee. (2017). Syringeless electrospinning toward versatile fabrication of nanofiber web, *Sci Rep*, **7**, pp. 41424.
61. He, H.-W., et al. (2016). Solvent-free thermocuring electrospinning to fabricate ultrathin polyurethane fibers with high conductivity by in situ polymerization of polyaniline, *RSC Adv*, **6**, pp. 106945–106950.
62. Han, D. and A.J. Steckl. (2019). Coaxial electrospinning formation of complex polymer fibers and their applications, *ChemPlusChem*, **84**, pp. 1453–1497.
63. Xu, C., et al. (2020). Polymer–mesoporous silica nanoparticle core–shell nanofibers as a dual-drug-delivery system for guided tissue regeneration, *ACS Appl Nano Mater*, **3**, pp. 1457–1467.
64. Huling, J., et al. (2020). Development of UV-reactive electrospinning method based on poly (ethylene glycol) diacrylate crosslinking, *Curr Dir Biomed Eng*, **6**, pp. 189–192.
65. Zhao, Y.-T., et al. (2020). Self-powered portable melt electrospinning for in situ wound dressing, *J Nanobiotechnol*, **18**, pp. 111.
66. Dabirian, F., et al. (2011). A comparative study of jet formation and nanofiber alignment in electrospinning and electrocentrifugal spinning systems, *J Electrostat*, **69**, pp. 540–546.
67. Valipouri, A., S.A.H. Ravandi, and A. Pishevar. (2013). A novel method for manufacturing nanofibers, *Fibers Polym*, **14**, pp. 941–949.
68. Gholipour-Kanani, A. (2022). A review on centrifugal and electro-centrifugal spinning as new methods of nanofibers fabrication, *J Text Polym*, **10**, pp. 41–55.
69. Liu, S.-L., et al. (2013). Assembly of oriented ultrafine polymer fibers by centrifugal electrospinning, *J Nanomater*, 2013, Article ID 713275, https://doi.org/10.1155/2013/713275.

70. Dabirian, F., S.A. Hosseini Ravandi, and A.R. Pishevar. (2013). The effects of operating parameters on the fabrication of polyacrylonitrile nanofibers in electro-centrifuge spinning, *Fibers Polym*, **14**, pp. 1497–1504.

71. Khamforoush, M., et al. (2014). The influences of collector diameter, spinneret rotational speed, voltage, and polymer concentration on the degree of nanofibers alignment generated by electrocentrifugal spinning method: modeling and optimization by response surface methodology, *Korean J Chem Eng*, **31**, pp. 1695–1706.

72. Hosseinian, H., et al. (2019). Determining the effect of centrifugal and electrical forces on the jet behaviors, the nanofiber structure, and morphology, *Polym Adv Technol*, **30**, pp. 941–950.

73. Kim, J.I., et al. (2016). A controlled design of aligned and random nanofibers for 3d bi-functionalized nerve conduits fabricated via a novel electrospinning set-up, *Sci Rep*, **6**, pp. 23761.

74. Sabzehmeidani, M.M., M. Ghaedi, and H. Karimi (2021), Chapter 10 - Photocatalytic activity based on electrospun nanofibers, in *Interface Science and Technology*, Ghaedi, M., ed., Elsevier, pp. 625–672.

75. Liu, W., S. Thomopoulos, and Y. Xia. (2012). Electrospun nanofibers for regenerative medicine, *Adv Healthc Mater*, **1**, pp. 10–25.

76. Xue, J., T. Wu, and Y. Xia. (2018). Perspective: aligned arrays of electrospun nanofibers for directing cell migration, *APL Mater*, **6**, pp. 120902.

77. Sabetzadeh, N. and A.A. Gharehaghaji. (2017). How porous nanofibers have enhanced the engineering of advanced materials: a review, *J Text Polym*, **5**, pp. 3–21.

78. Khajavi, R. and M. Abbasipour. (2012). Electrospinning as a versatile method for fabricating coreshell, hollow and porous nanofibers, *Sci Iran*, **19**, pp. 2029–2034.

79. Ji, L., et al. (2008). Preparation and characterization of silica nanoparticulate-polyacrylonitrile composite and porous nanofibers, *Nanotechnology*, **19**, pp. 085605.

80. Lv, J., et al. (2019). Nanofiber network with adjustable nanostructure controlled by PVP content for an excellent microwave absorption, *Sci Rep*, **9**, pp. 4271.

81. Lu, Z., et al. (2019). Synthesis of hollow nanofibers and application on detecting SF6 decomposing products, *Front Mater*, **6**, DOI: 10.3389/fmats.2019.00183.

82. Chiang, Y.-C., W.-T. Chin, and C.-C. Huang. (2021). The application of hollow carbon nanofibers prepared by electrospinning to carbon dioxide capture, *Polymers*, **13**, pp. 3275.

83. Mamun, A., T. Blachowicz, and L. Sabantina. (2021). Electrospun nanofiber mats for filtering applications: technology, structure and materials, *Polymers*, **13**, pp. 1368.

84. Ritz, B., B. Hoffmann, and A. Peters. (2019). The effects of fine dust, ozone, and nitrogen dioxide on health, *Dtsch Arztebl Int*, **51–52**, pp. 881–886.

85. Sharma, S. and A. Bhattacharya. (2017). Drinking water contamination and treatment techniques, *Appl Water Sci*, **7**, pp. 1043–1067.

86. Appert-Collin, J.-C. and D. Thomas (2017), 2: Fibrous media, in *Aerosol Filtration*, Thomas, D., et al., eds., Elsevier, pp. 31–47.

87. Li, Q., et al. (2016). An electrospun polycarbonate nanofibrous membrane for high efficiency particulate matter filtration, *RSC Adv*, **6**, pp. 65275–65281.

88. Tlili, I. and T.A. Alkanhal. (2019). Nanotechnology for water purification: electrospun nanofibrous membrane in water and wastewater treatment, *J Water Reuse Desalin*, **9**, pp. 232–248.

89. Olad, A., F. Farshi, and J. Ettehadi. (2012). Electrocatalytic reduction of nitrate ions from water using polyaniline nanofibers modified gold electrode, *Water Environ Res*, **84**, pp. 144–149.

90. Malani, A.S., A.D. Chaudhari, and R.U. Sambhe. (2015). A review on applications of nanotechnology in automotive industry, world academy of science, engineering and technology, *Int J Mech Mechatron Eng*, **10**, pp. 36–40.

91. Natrayan, L., et al. (2021). Processing and characterization of carbon nanofibre composites for automotive applications, *J Nanomater*, **2021**, pp. 7323885.

92. Kucukali-Ozturk, M., et al. (2017). Nanofiber-enhanced lightweight composite textiles for acoustic applications, *J Ind Text*, **46**, pp. 1498–1510.

93. Ulrich, T. and J.P. Arenas. (2020). Sound absorption of sustainable polymer nanofibrous thin membranes bonded to a bulk porous material, *Sustainability*, **12**, pp. 2361.

94. Jung, J. and H.A. Sodano. (2020). Aramid nanofiber reinforced rubber compounds for the application of tire tread with high abrasion resistance and fuel saving efficiency, *ACS Appl Polym Mater*, **2**, pp. 4874–4884.

95. Dziubak, T. (2019). Properties of material with nanofiber layer used for filtering the inlet air of internal combustion engines, *Combust Engines*, **177**, pp. 66–75.

96. Dziubak, T. (2021). Material properties analysis with addition of nanofibres for air intake filtration in internal combustion engines, *Int J Automot Mech Eng*, **18**, pp. 8621–8636.

97. Wu, H., et al. (2020). Review of application and innovation of geotextiles in geotechnical engineering, *Materials* (Basel), **13**, pp. 1774.

98. Balla, V.K., et al. (2019). Additive manufacturing of natural fiber reinforced polymer composites: processing and prospects, *Compos B Eng*, **174**, pp. 106956.

99. Lin, C., X. Zhang, and J. Han. (2019). Comprehensive material characterizations of pavement structure installed with wicking fabrics, *J Mater Civil Eng*, **31**, pp. 04018372.

100. Czigány, T., J. Vad, and K. Pölöskei. (2005). Basalt fiber as a reinforcement of polymer composites, *Period Polytech Mech Eng*, **49**, pp. 3–14.

101. Alshaaer, M. and H.Y. Jeon, Geopolymers and other geosynthetics. 2020: IntechOpen.

102. Zhong, W. (2016), 3: Nanofibres for medical textiles, in *Advances in Smart Medical Textiles*, van Langenhove, L., ed., Woodhead Publishing: Oxford, pp. 57–70.

103. Mahmoudi, N., et al. (2017). Temporary skin grafts based on hybrid graphene oxide-natural biopolymer nanofibers as effective wound healing substitutes: pre-clinical and pathological studies in animal models, *J Mater Sci Mater Med*, **28**, pp. 73.

104. Liu, X., et al. (2021). Electrospun medicated nanofibers for wound healing, *Membranes*, **11**, pp. 770.

105. Vasconcelos, A. and A. Cavaco-Paulo. (2011). Wound dressings for a proteolytic-rich environment, *Appl Microbiol Biotechnol*, **90**, pp. 445–60.

106. Azarniya, A., et al. (2019). Modification of bacterial cellulose/keratin nanofibrous mats by a tragacanth gum-conjugated hydrogel for wound healing, *Int J Biol Macromol*, **134**, pp. 280–289.

107. Zahedi, E., et al. (2019). Fabrication and characterization of core-shell electrospun fibrous mats containing medicinal herbs for wound healing and skin tissue engineering, *Mar Drugs*, **17**, pp. 27.

108. Samadian, H., et al. (2020). Electrospun cellulose acetate/gelatin nanofibrous wound dressing containing berberine for diabetic foot ulcer healing: in vitro and in vivo studies, *Sci Rep*, **10**, pp. 8312.

109. Dong, W.H., et al. (2020). Performance of polyvinyl pyrrolidone-isatis root antibacterial wound dressings produced in situ by handheld electrospinner, *Colloids Surf B*, **188**, pp. 110766.

110. Eskandarinia, A., et al. (2020). A novel bilayer wound dressing composed of a dense polyurethane/propolis membrane and a biodegradable polycaprolactone/gelatin nanofibrous scaffold, *Sci Rep*, **10**, pp. 3063.

111. Eslahi, N., et al. (2020). Processing and properties of nanofibrous bacterial cellulose-containing polymer composites: a review of recent advances for biomedical applications, *Polym Rev*, **60**, pp. 144–170.

112. Obregón, R., J. Ramón-Azcón, and S. Ahadian (2017), Chapter 19 - Nanofiber composites in blood vessel tissue engineering, in *Nanofiber Composites for Biomedical Applications*, Ramalingam, M. and S. Ramakrishna, eds., Woodhead Publishing, p. 483–506.

113. Karkan, S.F., et al. (2019). Electrospun nanofibers for the fabrication of engineered vascular grafts, *Journal of Biological Engineering*, **13**, pp. 83.

114. Pan, Y., et al. (2017). Small-diameter hybrid vascular grafts composed of polycaprolactone and polydioxanone fibers, *Sci Rep*, **7**, pp. 3615.

115. Xing, Y., et al. (2021). Gelatin coating promotes in situ endothelialization of electrospun polycaprolactone vascular grafts, *J Biomater Sci*, Polymer Edition, **32**, pp. 1161–1181.

116. Mazaheri, M., et al. (2015). Nanomedicine applications in orthopedic medicine: state of the art, *Int J Nanomed*, **10**, pp. 6039–6053.

117. Nedjari, S., F. Awaja, and G. Altankov. (2017). Three dimensional honeycomb patterned fibrinogen based nanofibers induce substantial osteogenic response of mesenchymal stem cells, *Sci Rep*, **7**, pp. 15947.

118. Samadian, H., et al. (2021). Bioengineered 3D nanocomposite based on gold nanoparticles and gelatin nanofibers for bone regeneration: in vitro and in vivo study, *Sci Rep*, **11**, pp. 13877.

119. Makris, E.A., et al. (2015). Repair and tissue engineering techniques for articular cartilage, *Nat Rev Rheumatol*, **11**, pp. 21–34.

120. Eslahi, N., M. Abdorahim, and A. Simchi. (2016). Smart polymeric hydrogels for cartilage tissue engineering: a review on the chemistry and biological functions, *Biomacromolecules*, **17**, pp. 3441–3463.

121. Rana, D., et al. (2017), 13: Nanofiber composites in cartilage tissue engineering, in *Nanofiber Composites for Biomedical Applications*, Ramalingam, M. and S. Ramakrishna, eds., Woodhead Publishing, pp. 325–344.

122. Leena, M.M., et al. (2021), 6: Nanofiber-integrated hydrogel as nanocomposites for tissue engineering, in *Bionanocomposites in Tissue Engineering and Regenerative Medicine*, Ahmed, S. and Annu, eds., Woodhead Publishing, pp. 119–147.

123. Liu, X., et al. (2020). HBC-nanofiber hydrogel scaffolds with 3D printed internal microchannels for enhanced cartilage differentiation, *J Mater Chem B*, **8**, pp. 6115–6127.

124. Sun, Y., et al. (2022). Bioinspired supramolecular nanofiber hydrogel through self-assembly of biphenyl-tripeptide for tissue engineering, *Bioactive Materials*, **8**, pp. 396–408.

125. Doblado, L.R., C. Martínez-Ramos, and M.M. Pradas. (2021). Biomaterials for neural tissue engineering, *Front Nanotechnol*, **3**, https://doi.org/10.3389/fnano.2021.643507.

126. Yao, S., et al. (2018). Hierarchically aligned fibrin nanofiber hydrogel accelerated axonal regrowth and locomotor function recovery in rat spinal cord injury, *Int J Nanomed*, **13**, pp. 2883–2895.

127. Steel, E.M., J.-Y. Azar, and H.G. Sundararaghavan. (2020). Electrospun hyaluronic acid-carbon nanotube nanofibers for neural engineering, *Materialia*, **9**, pp. 100581.

128. Garrudo, F.F.F., et al. (2019). Polyaniline-polycaprolactone blended nanofibers for neural cell culture, *Eur Polym J*, **117**, pp. 28–37.

129. Kamsani, N.H., et al. (2021). Biomedical application of responsive 'smart' electrospun nanofibers in drug delivery system: a minireview, *Arab J Chem*, **14**, pp. 103199.

130. Yu, D.G., et al. (2018). Electrospun amorphous solid dispersions of poorly water-soluble drugs: a review, *J Control Release*, **292**, pp. 91–110.

131. Li, F., et al. (2020). Stimuli-responsive nano-assemblies for remotely controlled drug delivery, *J Control Release*, **322**, pp. 566–592.

132. Mansouri Shirazi, N., N. Eslahi, and A. Gholipour-Kanani. (2021). Production and characterization of keratin/tragacanth gum nanohydrogels for drug delivery in medical textiles, *Front Mater*, **8**, https://doi.org/10.3389/fmats.2021.720385

133. Ghaffari, R., et al. (2018). Dual-sensitive hydrogel nanoparticles based on conjugated thermoresponsive copolymers and protein filaments for triggerable drug delivery, *ACS Appl Mater Interfaces*, **10**, pp. 19336–19346.

134. Puiggalí-Jou, A., et al. (2018). Smart drug delivery from electrospun fibers through electroresponsive polymeric nanoparticles, *ACS Appl Bio Mater*, **1**, pp. 1594–1605.

135. Amarjargal, A., et al. (2019). On-demand drug release from tailored blended electrospun nanofibers, *J Drug Deliv Sci Technol*, **52**, pp. 8–14.

136. Schoeller, J., et al. (2021). pH-responsive chitosan/alginate polyelectrolyte complexes on electrospun PLGA nanofibers for controlled drug release, *Nanomaterials*, **11**, pp. 1850.

137. Banerjee, A., et al. (2021). Magneto- and opto-stimuli responsive nanofibers as a controlled drug delivery system, *Nanotechnology*, **32**, doi: 10.1088/1361-6528/ac2700.

Chapter 2

Science and Applications of Polymeric Nanofibers

S. Chakraborty,[a] Pragati Bajpai,[a] L. Chakraborty,[b] and Shanu Prabhakar[a]

[a]*UPTTI Kanpur, Uttar Pradesh 208001, India*

[b]*Govt. Polytechnic, Nagpur, Maharashtra 440001, India*

supriyo1970@gmail.com

Polymeric nanofibers have immense potential and a large number of important applications such as in the biomedical field (such as tissue engineering and drug delivery), filtration, water, and air purification, energy storage, catalysis, sensors, and various environmental applications. Synthesis of nanofibers and such products for a particular purpose requires a lot of understanding of the physics and chemistry involved. Careful experimentation and a knowledge-based approach are the keys to making nanofibers of some particular use. In this chapter, various methods of synthesis of nanofibers have been discussed, with special stress on electrospinning technology, namely the physical principles involved in electrospinning. Other methods such as TIPS (thermally induced phase separation), self-

Handbook of Nanofibers and Nanocomposites: Characteristics, Synthesis, and Applications in Textiles
Edited by Mohd Yusuf and Aminoddin Haji

ISBN 978-981-4968-77-5 (Hardcover), 978-1-003-43274-6 (eBook)
www.jennystanford.com

assembly, or template synthesis have been also touched. Various application of these fibers has been reviewed with some details.

2.1 Introduction

There are several techniques where organic or inorganic polymeric materials can be produced to shrink/thinned from micrometers (e.g., 10–100 μm) to submicron or nanometers (e.g., 10^{-2}–10^{-1} μm). Polymeric nanofibers have many useful and remarkable characteristics as they have a very large surface-area-to-volume ratio. In terms of this factor, nanofibers show much superior to the micro denier fibers (by a factor of 10^3). Various functionalities can be introduced to the fibers, also they are superior in terms of mechanical properties (such as stiffness and tensile strength) compared to any other known form of the material [1]. These outstanding properties make the polymer nanofibers to be optimal candidates for many important applications.

2.2 Polymeric Materials for Nanofibers

Various polymers exist in nature, such as leather, cellulosic materials, woody substances, various types of grass, etc., has been used to produce papers, adhesives, composites, or used as textile fibers. The modified natural fibers were next to find their importance namely the cellulose nitrates found to attain commercial importance (a century before) as it was used in firm collars and cuffs. Cellulose nitrate was also used in motion picture films. The viscose rayon fibers were produced by dissolving cellulose as soda-cellulose xanthate in an alkaline solution and then spun into fine filaments by wet spinning in a coagulating bath. The viscose process is still in use today [2]. Starting from the first development of polymeric materials such as Bakelite that found application in various electrical appliances and phonograph records, various other purely synthetic or modified polymers have been widely used for the production of fibers or nanofibers for various applications.

A wide range of polymeric materials is used in making nanofibers by any of the various available techniques or very popularly and specifically by the electrospinning method. Natural polymeric materials such as collagen, cellulosic materials, silk, keratin, polysaccharides, chitosan, nucleic acids, and alginates can be named, and in the synthetic fiber group polycaprolactone, polyurethanes, poly(lactic-co-glycolic acid), poly(3-hydroxybutyrate co-3-hydroxy-valerate), etc. These polymeric materials have been used for nanofibers production of sub-micro to nano range that can be used for various applications. Various characterization techniques involved for the study of these nanofibers can be scanning electron microscopy (SEM), transmission electron microscopy (TEM), wide-angle X-ray diffraction (WAXS), small angle X-ray scattering (SAXS), environmental scanning electron microscopy, X-ray diffractometry, Fourier-transform IR (FTIR), energy dispersive X-ray, X-ray photoelectron spectroscopy, attenuated total reflectance IR spectroscopy, nuclear magnetic resonance, differential scanning calorimeter, attenuated total reflectance FTIR spectroscopy [3].

2.3 Production of Nanofibers: Various Techniques

Usually, there are various ways polymeric by which nanofibers can be synthesized, such as by using electrospinning techniques, STEP (spinneret-based tunable engineered parameters) or drawing techniques, template synthesis, phase separation/inversion, molecular self-assembly, air jet spinning, centrifugal spinning, interfacial polymerization of nanofibers [4]. Because of simplicity, cost-effectiveness, feasibility, flexibility to produce uninterrupted filaments from wide ranges of polymers, and the possibility of scaling up, electrospinning has attracted maximum attention and holds the position of most popular technology.

2.3.1 Drawing

Drawing is another method utilized to produce fibers. It is the drawing or stretching of the fiber that requires the polymeric material to have a pronounced viscoelastic nature to experience large deformations, and at the same time being enough cohesive to support the pulling stresses. This process is quite comparable to the dry spinning process, where long filaments are produced from polymeric solutions using appropriate stretch and simultaneous drawing to dry spinning. To note that only viscoelastic cohesive polymeric material can undergo such strong deformations. In addition, the pulling action always takes place with solidification that converts the feed material to a solid fiber. In the melt spinning process, solidification is achieved by cooling, while for a dry spinning process, evaporation of solvent takes place. This method is a rather simple and advantageous one, it requires only a sharp tip or a micropipette to produce nanofibers. A sharp tip is used to draw a droplet of a previously prepared polymer solution as a wet fiber. The solvent used for the preparation of polymer solution is evaporated due to increased surface area helps to achieve solidification. For example, very small (only a few ml) droplets of a solution can be deposited on a silica surface and allowed to be evaporated slowly. After waiting for a few minutes, the solution became more concentrated at the edge of the droplet attaining an appropriate concentration. A dipped micropipette can then be pulled away from the droplet at a certain speed (say 100 μm/s). A schematic representation of the technique is displayed in Fig. 2.1 [5]. Researchers achieved enhanced thermal conductivity nanofibers 20 times greater than the bulk fibers and close to electrospun webs [6].

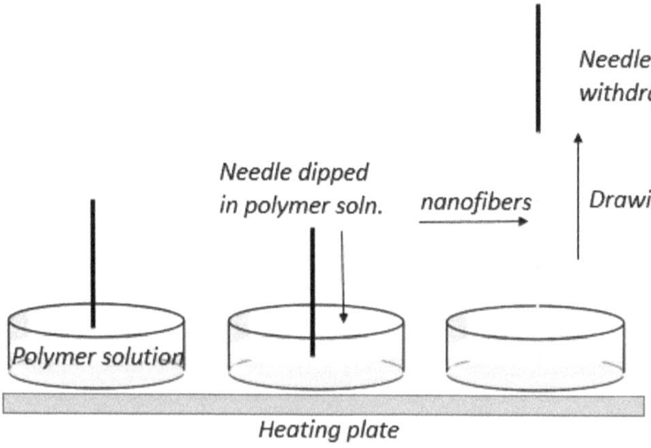

Figure 2.1 Simple schematic representation of the mechanism of production of nanofibers by drawing technique and uncoiling of the polymeric chains (polyethylene) in solvent [6].

2.3.2 Spinneret-Based Tunable Engineered Parameters

Controlled fabrication and deposition of smooth and uniform diameter nanoscale fibers of sizable lengths are still challenging as obtaining a high length-to-diameter ratio (L/D) to a great value (~2000 or more) at the submicron range makes the fiber very fragile and difficult to collect and process. The techniques mentioned have not yet been foolproof in this regard, which limits the use of such fibers. Production of nanofibers of a controlled diameter of nano range has been a challenge for technicians since its inception. The STEP (Fig. 2.2) techniques claim to produce precise nanofibers from an adequately concentrated polymer solution on a cylindrical surface with regular geometric spacing, which can be explored for designing novel sensors [7].

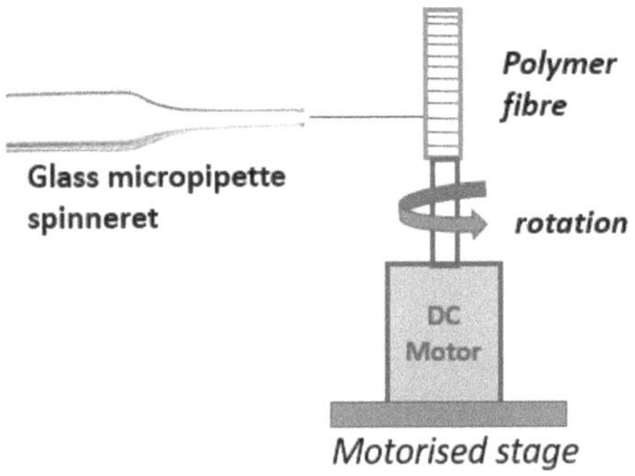

Figure 2.2 Schematic representation of the setup for producing tuneable uniform diameter from polymeric solution [7].

2.3.3 Template Synthesis Method of Producing Nanofibers

Apart from the few methods mentioned there are numerous chemical approaches to preparing nanofibrous or nanoporous materials. Nanoporous membrane-type material is used in template synthesis for producing and desired quality of the material. Membranes involved in the production of nanofibers usually have cylindrical nanopores of the same diameter (Figs. 2.3 and 2.4). In essence, we view each of these pores as a beaker in which a piece of the desired material is synthesized. The pores in the template are of defined cylindrical shape that results in nanocylinders of the desired material from each pore. Nanocylindrical fibrils can be solid (fibrils) or hollow (tubules) that depends on the material processed and the chemical properties of the pore wall used [8].

Conducting polymer nanostructures can be produced using a soft template or hard template methods. The soft template technique is based on molecular self-assembly whereas hard templates replicate prevailing nanostructure through physical and chemical interactions. Synthesis of polyaniline granules, nanofibers, nanosheets, and other structures has been reported [9].

Synthesis of nanostructures including various fiber template methods can be simple and universal. This is a method for fabricating innovative conducting polymer nanostructures and composites in a controlled process. In addition, it is possible to design and synthesize fine mesostructured conducting polymers that find application in high-performance energy storage devices.

The nanopores on membranes used in the template usually are very uniform in size, and the diameter and the aspect ratio of the nanofibers using the template technique can be accurately controlled.

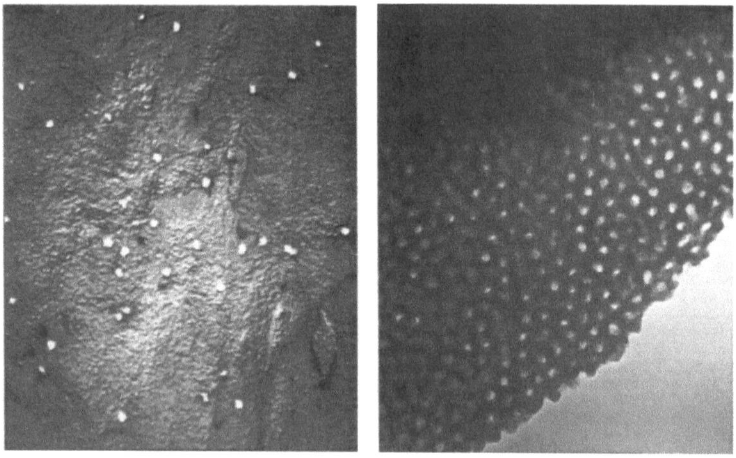

Figure 2.3 Nanoporous polycarbonate and alumina membranes used for synthesis of electronically conductive polymer nanostructures. Reprinted with permission from Ref. [8], Copyright 1995, American Chemical Society.

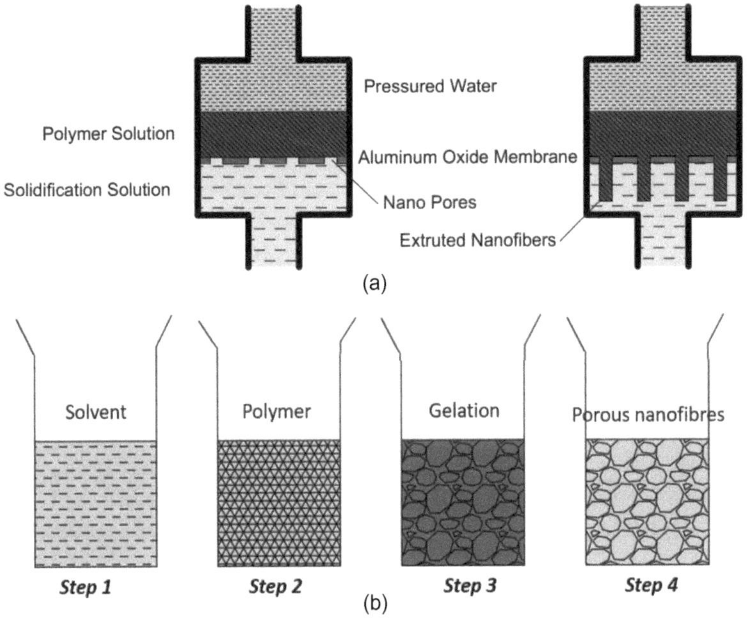

Figure 2.4 Synthetic nanoscale fibrous material preparation by (a) Template synthesis method and (b) Self-assembly method.

Another method includes the preparation of micro and nanostructured material with the aid of the template method in the chemical vapor deposition (CVD) process. The CVD technique is a very useful methodology, which involves gas molecules being decomposed into reactive species that result in nanofilm or particle growth. In some research approaches CVD technique and template synthesis methods have been combined to fabricate TiS_2-coated gold fibers for the preparation of electrodes for battery [10].

2.3.4 Thermally Induced Phase Separation

Nanofibrous scaffolds can be prepared by phase separation methods which is another exciting technique to produce nanofibers for biomedical applications. The phase separation technique typically includes five steps: dissolution of the target polymer in a proper solvent to attain a homogenized mixture, formation of gel, extraction by another solvent, freezing followed by freeze-drying, and ultimately fabricated porous nanofibrous foam (Fig. 2.4). This method may be combined with some other assembly techniques for controlling its final 3-dimensionsal structure but the disadvantage is that there is little control over fiber diameter and orientation, as well as poor mechanical properties. However, making a porous material of interest by phase separation method takes a lengthy method thus making it a time-consuming process [11, 12].

The thermally induced phase separation (TIPS) technique is related to the thermodynamic process involving the separation of a homogeneous mixture into multiple phases due to physical inconsistency. At a certain temperature region, the homogeneous polymer solution will become unstable thermodynamically (Fig. 2.5) and subsequently, phase separation will take place. A polymer-rich phase and a polymer-poor phase will be generated. Removal of the solvent by say freeze-drying, the polymer-rich phase will develop a 3D structure same time the polymer-deficient phase spaces will be generating the voids.

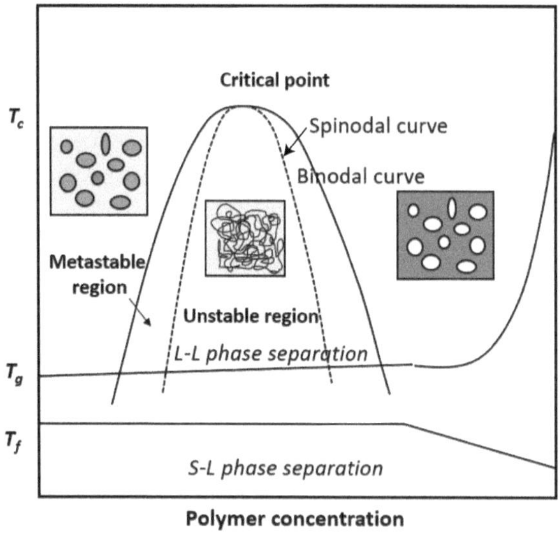

Figure 2.5 Schematic representation of the thermally induced phase separation process [13].

All the steps mentioned earlier namely raw material dissolution, gelation, solvent extraction, freeze-drying, etc. have a significant influence on the characteristics of the generated scaffold. Using this method the morphological structure of nanofibrous scaffolds can be controlled by altering various process and composition parameters, such as polymer concentration, the temperature of phase separation, solvent non-solvent exchange, and time [14]. To add, thermally induced phase separation can be coupled with a few additional processing techniques such as particulate leaching, and can generate scaffolds with complicated porous assemblies with well well-defined pore morphologies [15]. For example, sodium chloride particles with thicknesses of 200 ~ 450 μm complied with a warm poly(lactic-co-glycolic acid)/in tetrahydrofuran solution followed by cooling to a suitable gelation temperature. The gel complex formed in this case can be removed by using cold ethanol, the solvent was removed by washing with distilled water and leaching out the salt. Subsequently, after freeze-dried samples, a nanofibrous scaffold with a macro-porous structure was obtained. A 3D macro-porous scaffold of nanofibrous poly(L-lactic acid) was produced by using the phase separation method to improve functions such as mass transport and spreading of cells. As compared with solid pore PLLA scaffoldings, nanofiber-based scaffolds permitted an about two-time increase in the adhesion of osteoblast cells [16].

Thus, the TIPS technique is a promising way of producing nanofibrous scaffolds for biological applications having precise pore structure and dimension. Comparing various processing techniques for producing nanofibers shows that among drawing, template synthesis, phase separation, self-assembly, electrospinning methods can be used in the laboratory but in terms of options of scaling up and control of fiber dimensions or convenience of processing electrospinning techniques have the maximum potential [12, 17].

2.3.5 Polymer Nanofibers by Molecular Self-Assembly

It is the spontaneous organization of molecules in some structurally defined positioning or organizing through bonds such as van der Waals force of attraction or hydrogen bonds, electrostatic or hydrophilic attractions. A wide variety of self-assembled systems are available universally in various complex biological structures. Molecular self-assembly is a bottom-up approach that is used to create nanofibers from small building blocks such as molecules, peptides, or nucleic acids. Synthesis of nanostructured material is associated with the broad area of biology and nanochemistry and relates itself to various areas of study such as electrochemistry, molecular recognition, polymer science, colloid science, and advanced sophisticated characterization methods such as scanning probe microscopy [18, 19].

For molecular self-assembly components of supramolecular nanofibers can be selected widely from organic, polymeric, inorganic, and biomolecules. Organic constituents may consist of molecules, polymers, surfactants, synthetic amphiphiles, antibiotics, liquid crystals, and metal complexes; biomaterials may consist of amino acids, peptides, proteins, nucleotides, carbohydrates, polysaccharides, lipids; inorganic components include nanoparticles, nanorods, nanowires, nanoplates, nanosheets. Nanofibers can be self-assembled from synthetic appropriately designed amphiphiles in aqueous or in organic environments [20]. Polymer nanofibers as well as some extensive variety of mesoscopic scale structures such as vesicles, tubes, disks, lamellas, and helices are formed from suitably designed amphiphiles—of various types, such as bola amphiphiles and Gemini surfactants [21–24]. These amphiboles contain a polar head group and one or more hydrophobic tails.

Self-assembly processes of peptide-based nanofibers include many driving forces (*viz,* hydrophobic hydrophilic interactions, electrostatic force, hydrogen bond, π–π interaction, van der Waals forces, etc.) [25, 26], at the same time can be affected by different prevailing conditions such as pH of the system, ionic strength, and assembling rates [27]. Besides, multiple nanostructures such as twisted ribbons, helical ribbons, and nanotubes can coexist and transform with time [28].

Numerous self-assembling soft materials gathered a lot of attention such as molecular surfactants or coil/coil block copolymers, and molecular rods with flexible coils gained attention as self-assembler [29]. Molecular rod system has the distinct feature of configuration and shapes differences due to rigid and flexible parts, as well as amphiphilicity due to specific block compositions. Such molecules with rod and coil segments are able to show microphase separation in the ordered structure. As compared to the traditional coil–coil block copolymers, improved segregation results in an assembled structure with relatively smaller dimensions (a few nm). Different molecular rod blocks, such as helical peptides [30], isocyanates [31], mesogenic phenyl units, and rigid conjugate aromatic units [32], have been studied by researchers for supramolecular structures.

2.3.6 Electroconducting Nanofibers: Interfacial Polymerization

Aniline is converted to electroconductive polyaniline using an oxidant and a strong acid as a dopant [33]. Polyaniline nanostructures namely nanotubes, fibers, wires, and rods can be produced by chemical methods that require the addition of structure-promoting materials during polymerization. Zeolite channels have been reported to be used as templates to produce polyaniline nanostructures [34], and some other structures such as nanoporous membranes, functional surfactant molecules, polyelectrolytes, or organic dopants can be used. Polyaniline polymerization is carried out by the interfacial polymerization method, as the polymerization of aniline occurs at the interface between the two immiscible liquids. Aniline is dissolved in the organic phase and the aqueous phase contains the oxidant as a dopant. Polyaniline produced diffuses through the interface to the aqueous layer, and eventually fills it. PAni nanofibers can be found to have applications as sensors [35].

2.3.7 Electrospinning of Nanofibers

Of the various nanofibers' fabrication methods, the most convenient is electrostatic spinning or electrospinning, which is up scalable technique for the production of nanofibers. Currently, in many industrial applications, electrospinning technology can be found to be utilized. The earliest document of the accounts of electrostatic spinning of a polymer solution into nanofibers can be found way back in 1902. Since then, a large number of variations and improvements have been carried out over the last hundred years. Minimum components of the electrostatic spinning setup may include a needle that supplies the polymer solution at the desired rate, DC voltage applied to the electrode, and a collector (oppositely charged) to collect the nanofibers.

2.3.7.1 Components of the electrospinning process

A horizontal electrospinning machine is shown in Fig. 2.6 which shows the basic three components of the electrospinning machine essential to fulfill the production of a two-dimensional nanofibrous web in the process. These are (a) high-voltage (kV) supplier, (b) needle (~0.83 mm), and (c) drum collector for collecting the nanofibrous web. High electrical charge or electrostatic force in polymer solution is usually generated by a high voltage (5–50 kV in the process) so, the viscose polymer comes out from the needle tip and forms a prolated Taylor cone for the continuous formation of nanofibers. The positive electrode is placed into the solution needle tip and the other negative electrode is attached to the rotating drum collector.

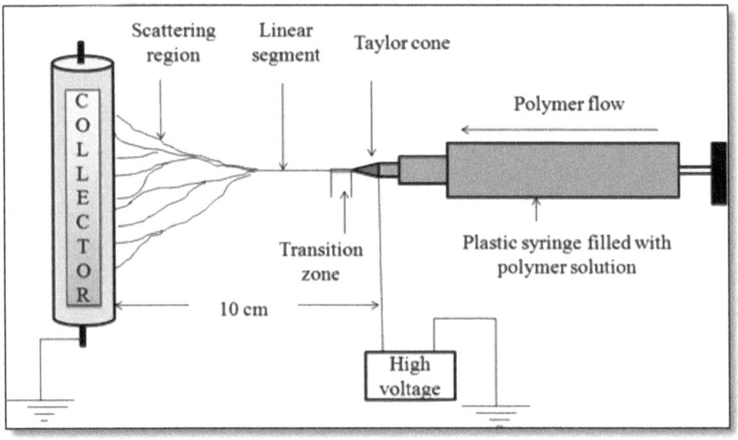

Figure 2.6 Schematic diagram of horizontal needle electrospinning machine.

2.3.7.2 Different types of nozzles

There are various types of nozzles used for producing nanofibers by electrospinning they are single-nozzle, multinozzle, co-spinning nozzle, etc. The electrospinning setup using a coaxial nozzle is shown in the following Fig. 2.7.

For the generation of the hollow nanofiber or core-sheath fibers, core-shell nozzle or coaxial spinnerets are used is usually used [36, 37]. The core-shell nozzle is made up of Teflon and stainless steel displayed in Fig. 2.7, needles are made with precision for the generation of the good quality of hollow fibers. Core-shell fibers are consisting of two layers of structure in the core of one material and the outer shell of another material. Both the outer and core layers can be retained and one of the components can be removed to produce a hollow or nanorod. The core-shell type mechanism produces fibers with diverse functionality that finds application in various innovative application areas such as ultrafiltration, tissue engineering, membranes, and drug delivery [38].

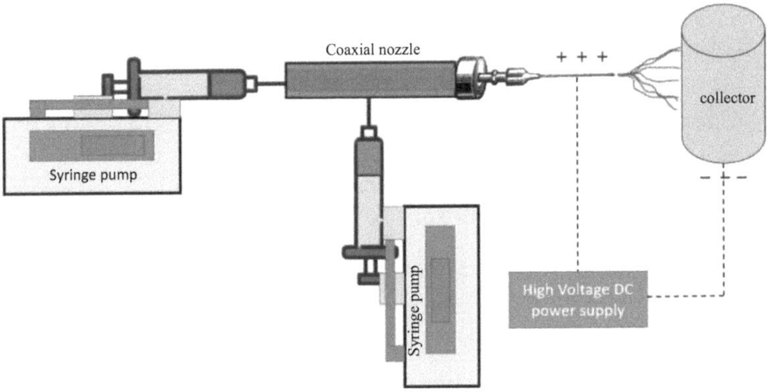

Figure 2.7 Schematic diagram of coaxial electrospinning for hollow or other nanofibers.

2.3.7.2.1 *Single nozzle*

The single nozzle technique utilizes the electrostatic force in the electrospun fabrication device to form a nanofibrous web from a viscous solution in nano- or microsize. A single nozzle is most commonly used to make solid two-dimensional nanofibrous or microfiber web. Single nozzle syringes come with different sizes of needles syringe.

2.3.7.2.2 *Multinozzle*

Multinozzle needles are used for high throughput/productivity of fiber. Such multi-nozzles are made with the bass material and Teflon which is restive to various corrosion effects caused by chemicals. This electrospinning technique is also capable of generating a higher electrostatic force than usual single-needle electrospinning.

2.3.7.2.3 *Co-spinning nozzle*

For the generation of micro or nanofiber with a combination of two different polymers or solvents. Such a co-electrospinning process is also suitable for a polymer couple which generates a precipitate while mixing different polymers [38].

2.3.7.3 Needleless spinneret

There are some other innovative techniques that serve the purpose of generating jets for electrospinning. Some examples can be stated as cylinder spinnerets, ball spinnerets, disk spinnerets, beaded chain spinnerets, spinal coil, cone spinneret, etc. those fall in the subcategory of rotary spinneret of the needleless spinneret group. Some examples of needless static spinneret are solution bubble type, plate spinneret, etc. [39].

2.3.7.4 Electrospinning setups and different collector types

Collectors for electrospun webs can also be categorized as rotary and static collectors. There are various types of collectors popularly used for collecting electrospun webs, such as drum roller collectors, disk collectors, rod collectors, liquid collectors, window collectors, and static collectors, including parallel ring collectors, an array of counter electrodes, etc. to name a few.

2.4 Material and Process Variables in Electrospinning

2.4.1 Material and Process Parameters Affecting Electrospinning

Polymers are to be used for spinning electrospun fibers for diverse applications, hence merits of different raw material characteristics must be considered critically for respective requirements. Nanofibers spun can be produced of various functional and morphological properties by varying process and material parameters. Employing some suitable technologies nanocomposites can be manufactured directly by spinning or nanofibers may be used as a reinforcing material in a matrix of other materials. To produce ceramic nanofibers, nanofibers are required to be post-processed after electrospinning. Hence an elementary understanding of the different groups of material is important before choosing to produce the desired nanofiber.

Polymeric electrospun nanofiber production and functional properties are affected by many parameters related to the polymer properties such as its polymer molecular constitution (e.g., polyethylene oxide, PEO, collagen polyethylene oxide, poly(D,L-lactic acid), PDLA, etc.,) whereas molecular weight, viscoelastic properties, surface tension, the electrical conductivity of a solution, the dielectric effect of the solvent, type of solvent used are going to have telling effect on the spinning process as well as properties of the fibers. Process variables include applied voltage, throughput rate, temperature, the distance between tip and collector, etc. Environmental variables such as relative humidity, temperature, pressure, etc. also affect the spinning process and resulting properties of the fibers.

2.4.1.1 Electrospinning: Effect of material variables

The polymer's molecular structure has a significant impact on its solubility in the solvent. In general, a polymer sample with a high molecular weight will be less soluble and take a longer time to get dissolved in the same solvent than a polymer sample with a low molecular weight. Longer chain molecules have higher intermolecular interactions, and solvent molecules take longer to permeate into the polymer bulk. Because the covalent connection between the individual polymer molecules is substantially stronger than the secondary forces exerted by polymer-solvent interactions, cross-linked polymers do not disintegrate. From the understanding of the solvent and solute molecular interactions, suitable strategies can be set to derive special nanofiber properties.

Commencement of electrospinning involves the charged solution to overcome the **surface tension**, (surface tension being responsible for the formation of beads) that play important role in the dynamic behavior of the electrospun jets. During the electrospun fiber spinning, high **viscosity** (meaning a higher level of interaction of polymer and solvent) will help reduce the effect of surface tension. All different polymers with various concentrations for different solvents may require unique process parameters to produce the desired quality of fibers. Hence higher viscosity can be expected to increase the diameter of the spun filament whereas higher surface tension will result in more beads. An increase in the electrical conductivity of the polymer solution results in finer fibers being produced.

Electrospinning can be done with either molten polymers or using polymer solutions. The form and characteristics of the resulting fibers are heavily impacted by the polymer solution's properties. The droplet formed at the tip of the needle is attenuated as a cone during electrospinning, resulting in exceedingly tiny fibers. The extent of stretching of the solution in the form of a filament is determined by its electrical characteristics (electrical conductivity and dielectric properties), solution viscosity, and surface tension. The speed at which the solvent evaporates influences the viscosity of the polymer solution being strained. The viscosity of a solution can be determined by the solubility of the polymer in the solvent, as well as the types of polymers that can be mixed.

2.4.2 Effect of Process Variables

Various process variables include *applied voltage, flow rate, spinneret to collector distance*, etc. Increased applied voltage results in finer diameter of filaments, but not necessarily always, as at an increased level it may affect the increase in diameter, too. An increase in mass throughput rate or feed rate fiber diameter can increase and a beaded structure may result when the feed rate is too high. With the increase in a spinneret to collector distance fiber diameter decreases in the other way which may result in a beaded structure. The ambient condition also affects fiber quality in electrospinning. Such as at lower temperatures fiber diameter and viscosity decrease. Increased humidity may result in the generation of circular pores on the fibers. It is noted that material variables related to polymer or its solubility/concentration, solvent properties, etc. are more important and related to the successful spinning of the required nanofibers.

2.4.3 Electrospinning Technology: Mathematical Modeling

The electrospinning method is a microscale and nanosized fiber synthesis technique that uses a polymeric solution or a polymer melt and an applied DC electric power. Electrospinning technologies require both the science and methodology of electrospraying as well as traditional dry spinning procedures to produce fibrous webs but without coagulation steps associated with solution spinning [40].

A polymeric flow (solution or melt) that is highly electrically charged is made to form uniform drops of size about 0.1 mm in diameter usually produced by enforcing a potential of 5–10 kV AC or DC to liquids in small capillaries [41]. This topic has been discussed and researched over the past many years and has been found to have difficulties associated with flow rate limitations.

The minimum equipment requirements for the demonstration of simple electrospinning in the laboratory are as follows:

1. Viscous polymer solution or a melt
2. Hollow tubular element that is used to act as an electrode, maintained in contact with the polymer solution
3. A high-voltage DC source connected to the mentioned electrode
4. A properly grounded or oppositely charged static or rotating surface that collects the nanofibers

Typically, in an electrospinning experimental setup, a very small volume of polymeric solution (a polymer of interest is dissolved in suitable a solvent) is prolated to a very fine filament by the application of an electrostatic field. Adequate viscosity of the polymer solution prevents drippage from the pipette tip by the action of gravity. Various physical and material parameters and factors including polymer type, solution viscosity, applied electrostatic field, and feed rate affects the quality and functionality of the produced nanowebs.

Primarily during the electrospinning of nanofibers, visibly from a liquid polymeric droplet, a jet is ejected out from the meniscus. The liquid meniscus formed at the tip of the capillary form in the shape of a spherical cap may have a capillary stress of an order of γ/r, where γ is the surface tension of the polymer solution, and r is the principal curvature of the meniscus. It is a known fact that applying a static electric field to a liquid droplet causes instability of the liquid shape, subsequent deformation occurs to the droplet or the meniscus, and finally an ejection of liquid filaments and/or fine droplets [42, 43].

In the late 1960s, Taylor first explained the relationship of liquid meniscus shape and applied DC power, subsequently finding the equilibrium droplet conical shape, which came to be known as the Taylor cone [43, 44].

The steady-state liquid droplet shape results from the balance of the electric forces and the surface tension, regardless of any viscoelastic forces that vanish in static conditions. A liquid droplet in a static condition attached to a positive electrode located at some distance from an opposite electrode can be considered a perfect ionic conductor, even if it is a leaky dielectric liquid [43–45]. The mentioned phenomenon is due to the interaction between Coulomb repulsion of similar charges, which promotes droplet distortion and partitioning, and the surface tension, which opposes droplet division. As the charge inside the fluid reaches a critical value, the liquid jet breaks out from the droplets from the needle tip. The electrospun jet so produced will move toward a lower potential, commonly which is a grounded collector. In general, more than 6kV of applied voltage has been found to distort the liquid droplet to take up the shape of a cone at the time of jet initiation [46].

Geoffrey Taylor's in the sixties presented a mathematical derivation of a model representing the nature of droplets placed in an electric from the fundamental understanding of the facts involved. A small electrically conductive liquid droplet is acted upon by an electric field, and the electric field causes the charged droplet to change its shape. The effect becomes more prominent as the applied voltage is increased up to a maximum. The magnitude of the force due to the applied electrical field gradually equals the forces due to surface tension and a cone shape starts to begin that has convex sides and a rounded tip. This ultimately approaches a conical shape that has a tip angle of 98.6°.

In a conductive liquid droplet subjected to an applied electrostatic field (in the condition that charge accumulates on the surface of the droplet and no electric field inside) the pressure balance is given in terms of surface tension, droplet radius, the total charge on droplet and permittivity of the vacuum by the following relation, where e is the total charge on the droplet, R is droplet radius, σ is the surface tension and ε_0 is the permittivity of vacuum (Eq. 2.1):

$$\Delta P = \frac{2\sigma}{R} - \frac{e^2}{32\pi^2\varepsilon_0 R^4} \tag{2.1}$$

It can be noted that the electrostatic pressure becomes dominant as the radius of the droplet becomes smaller or the charge density increases [47]. Taylor (1964) and Sherwood (1988) [48] found experimentally as well theoretically that a fluid droplet without any charge can become unsteady as (Eq. 2.2),

$$E \sim \sqrt{\frac{\gamma}{\varepsilon_0 \cdot r}} \tag{2.2}$$

where E is critical electric field stress in V/m, r is the radius of the drop before deformation, γ the surface tension of the fluid in N/m. Understanding the development of the unstable jet from the commencement of the instability and its behavior until the jet strikes the collector is a fundamental difficulty in the subject of electrically propelled jets.

2.4.3.1 Bending instability

The bending or whipping motion rapidly gets thinner and the jet breaks it into a spray of polydispersed droplets with sizes of the order of tens of microns. In the case of highly charged jets, the electric repulsive forces overcome surface tension, causing the portion of the jet to buckle or bulge in various locations, from which finer sub-jets are released. This split-up type of jet is associated with the electrostatic Rayleigh breakup mode [47].

Yarin et al. [48] developed a viscoelastic model for the electrified jet to mathematically analyze the bending unstableness of the electrospun fluid jet. This model for the electrospinning process is supported by an experimental setup where they spun using an aqueous solution of polyethylene oxide. During spinning an electrically influenced bending instability occurred, which is caused by disturbances of the lateral positions and the lateral movement of the jet growth. All the points considered along the electrospun filament carry the same charge a repulsive force between all the same charges with the jet causes each small portion of the jet to elongate continuously along a fluctuating path till the jet gets solidified. Yarin et al. [48] showed the bending instability of the electrified jet by taking a high-speed photograph. It was observed that about 20 ms after a loop had first formed, a new set of instability is occurring with a time interval of less than a millisecond, producing thinner and curved loops. Instability develops when the viscoelastic stress in the jet is sufficiently relaxed. Subsequently, they develop small spirals that coil around the first loop. The bending and spiraling filament movement of the filament can be conceived as shown in Fig. 2.10.

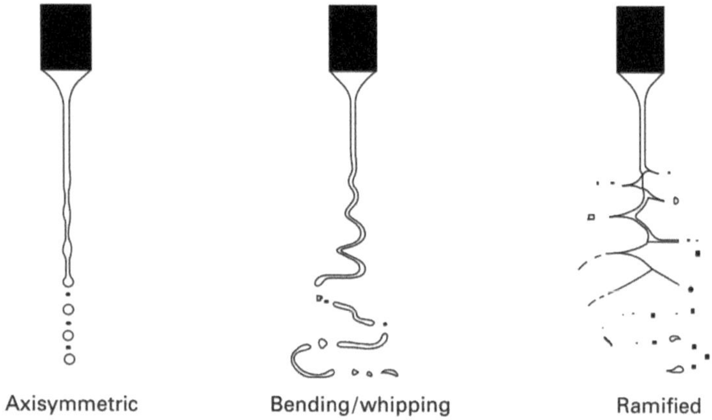

Figure 2.8 Schematic representation of the jet breakup motion [70].

Yarin et al. [48] represented a small segment of the electrospun jet by a spring and dashpot (Fig. 2.8). Subsequently they solved the equations of mass balance, momentum balance, and Coulomb force and viscoelastic force balance and explained bending instability based on the model. As represented in Fig. 2.9 points A and B are of mass m_A m_B and possess electric charge e. If point A is fixed, then the Coulomb force of repulsion acting on point B will be $-e^2/l^2$. For the spring and dashpot unit AB which for a Maxwellian viscoelastic unit the corresponding stress which is pulling back B is given by (Eq. 2.3),

$$\frac{d\sigma}{dt} = E\frac{dl}{dt} - \frac{E}{\mu}\sigma \tag{2.3}$$

where σ is stress, l is the instantaneous length of the jet in figure (distance of point B from point A), E the elastic modulus, and μ the viscosity.

The momentum balance of point B is represented by the following equation (Eq. 2.4),

$$m\frac{dv}{dt} = -\frac{e^2}{l^2} - \frac{eV_0}{h} + \pi a^2\sigma \tag{2.4}$$

where a is the radius of the filament, v is the velocity of bead B, which satisfies the momentum balance equation (Eq. 2.5),

$$\frac{dl}{dt} = -v \tag{2.5}$$

The bending instability of the electrospun jet which occurs during its formation is an example of general instability of charges under electrical forces and in relation to the electrospinning mechanism, it can be explained as per Yarin et al. after James [49].

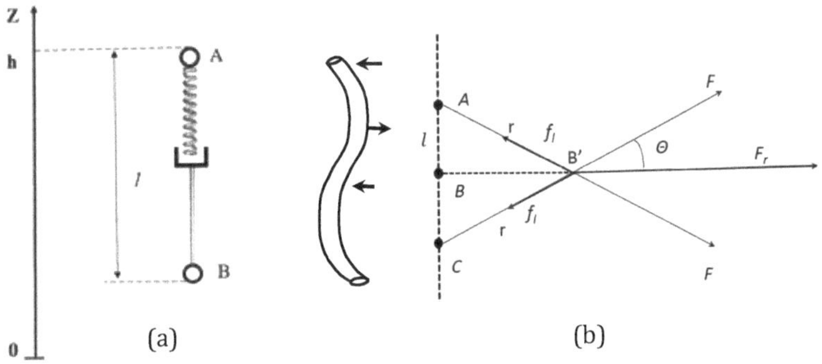

Figure 2.9 (a) Two imaginary nodes in an electrospinning filament constituted of a viscoelastic spring-dashpot arrangement. (b) Electrostatic repulsion acting on a viscoelastic unit.

It has been explained in the following way: considering three point-like charges, each having an electric charge of value e coulombs, originally all the three points A, B, and C are in a straight line as shown in Fig. 2.9. There are two Coulomb repulsive forces of magnitude $F = e^2/r^2$ and this repulsive force pushes point B in the direction of $F1$. When a perturbation causes the point B to move off the straight line by a small distance δ to another point, a net repulsive force $F_1 = 2 \cdot \cos\theta = (2e^2/r^3) \cdot \delta$ will work on the point charge B in a perpendicular direction, to the vertical line AC made by the fixed point of charges at point A and point C, in the direction of the disturbance, it tends to cause B to be deflected away. If the development of a small bending perturbation that can be considered as 'δ,' force balance will follow the (Eq. 2.6) given equation,

$$m\frac{d^2\delta}{dt^2} = \frac{2e^2}{l_1^3}\delta \tag{2.6}$$

Figure 2.10 Illustration showing the end of the straight segment, followed by the onset and development of three successive generations of right-handed bending instabilities. Reprinted with permission from Ref. [50], Copyright 2006, American Chemical Society.

The solution of the above equation indicates that a small perturbation created increases exponentially. In case the three points A, B, and C are connected the liquid jet force accompanying the liquid will attempt to counter the instability caused by the Coulomb repulsive forces.

Force balance equation can be written, recalling Newton's Second Law of motion as follows (Eq. 2.7),

$$m\frac{\partial^2 r_i}{\partial t^2} = \Sigma f_i$$
$$\Sigma f_i = f_C + f_E + f_V + f_S + f_A + f_G + \ldots \tag{2.7}$$

where m is the mass and the various forces are summed as Coulombic (with subscript C), electric field (with subscript E), viscoelastic (with subscript V), surface tension (with subscript S), air drag (with subscript A), gravitational (with subscript G).

In fact, surface tension forces in one of the most important factors in hydrodynamic jets that causes the jet to become unstable as these forces tend to reduce surface energy. In electrospinning, surface tension works against the inherent instability, caused by the electrostatic repulsion of different points, because bending is always associated with the increase of energy. The surface tension force acts on the ith bead and tends to restore the rectilinear shape of the bending part of the jet. It can be calculated from the surface tension coefficient, average radius, and curvature of the point considered from the coordinates of the beads [48]. In addition to this air drag and gravitational forces were also to be taken into consideration. To observe the trajectory of the liquid jet and its instability spatial and temporal perturbation was introduced and all equations were solved numerically.

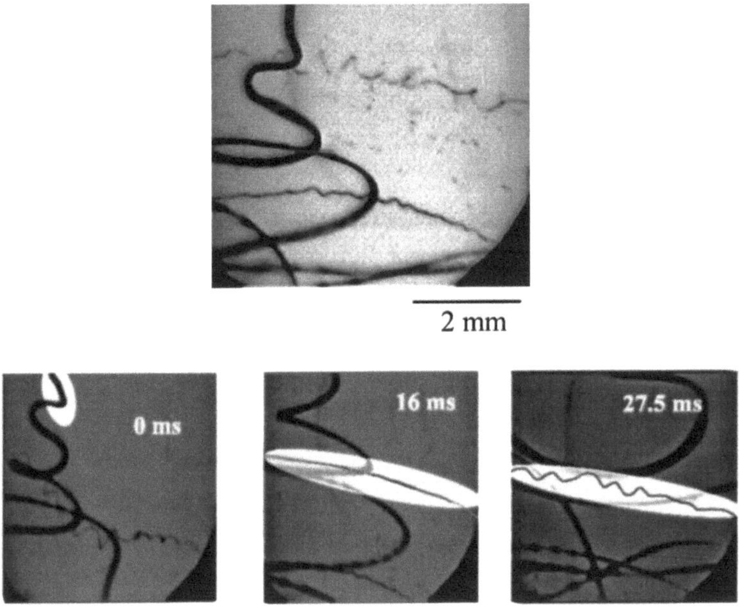

Figure 2.11 Sequentially taken photographic images showing the gradual development of instability with time during electrospinning. Reproduced from Ref. [48], with the permission of AIP Publishing.

From a series of images taken by an electronic camera a selected portion/segment of the filament was highlighted in each picture to show the extension of the filament and its instability. A rate of the extension was observed to 22 mm/s in this case and after a time gap of 22 ms slightly curved filament was observed. This continues to bend and more curved as shown in (Fig. 2.11) up to 27.5 ms. The smaller bent portions continue to be elongated, which is not visible as long as the charge on the jet produces enough force.

2.5 Applications of Polymeric Nanofibers

Polymeric nanofibers have numerous applications in various fields due to their unique properties and high surface-to-weight ratio. Nanofibers could be used in the biomedical field including tissue engineering, drug release, and wound dressing. In addition to the biomedical area due to the porous structures and large specific surface area, they have also found applications in the areas such as filtration, protective garments, energy storage, nanocomposites, environment, and nanosensors (Fig. 2.12).

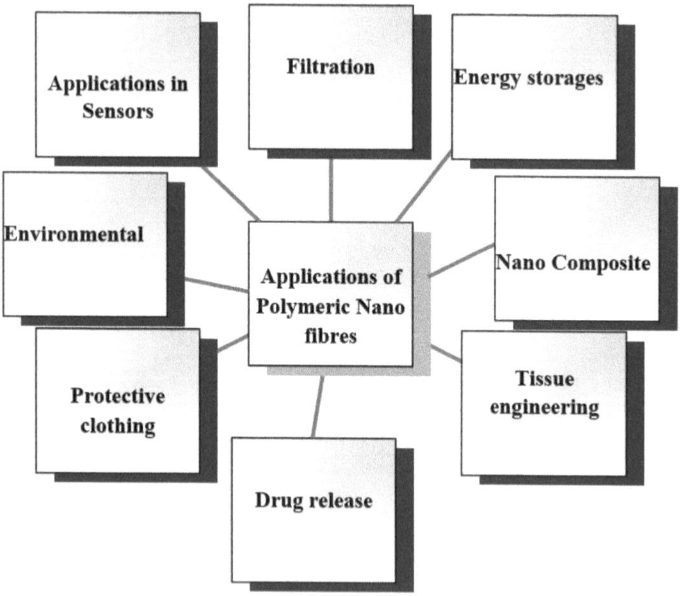

Figure 2.12 Applications of polymeric nanofibers.

2.5.1 Filtration

Filtration is a method of separation. Highly efficient filtering media with the ability to capture nanosized particles are widely employed in many technical domains. The global industrial filtration market size is rapidly growing to eradicate air and water pollution. Electrospinning can produce micro- and nanofibers, for a broad array of applications. For filtering, numerous polymeric nanofibers have been explored, like polyvinylidene fluoride nanofibers, which were electrospun into membranes and examined to see how structural factors relate to membrane separation properties. Electrospun membranes having properties similar to conventional microfiltration membranes were characterized [51].

Electrospun membranes are used in a variety of water purification applications, including membrane distillation and treatment before feeding or in nanofiltration techniques to remove metal ions, grease, and other pollutants [52]. Filter media made of fibrous materials with excellent filtration efficiency with less air resistance [53]. Since the size and structure of pathways of a filter must be identical in size to droplets or particles to be collected in the filter, electrospinning has a high ability to produce fiber media that can remove nanosized foreign particles efficiently. Using nanosized fibers in the filter is one effective method for producing extremely effective and efficient filter media [54].

Due to their high specific properties, low resistance, and better filtering efficacy, nanofibers are an intriguing material for numerous fields of applications, including healthcare, energy, and air filtration. Filters made of functional nanofibers can be used instead of glass fibers and charcoal. Non-selective filtering can be improved with these filters [55]. Antibacterial chemicals, such as silver (Ag), can be combined with these polymeric nanofibers to give the filters antimicrobial capabilities. Polyacrylonitrile (PAN), polyvinyl chloride (PVC), and cellulose acetate (CA) nanofiber membranes containing Ag nanoparticles were found to have antibacterial action against *E. coli* and *P. aeruginosa*, according to Neeta et al. [56].

Another polymer polysulfone is used for the removal of micro-particles from wastewater, increasing the life of ultrafiltration or nanofiltration membranes [57].

Nanofiber membranes made from specialized polymers can be utilized as molecular filters. For serving the more typical filtration purpose, chemical and biological weapon agents can also be detected and filtered using such filters [58].

A novel type of CNF-based fibrous membrane that can be able to filter and separate nanoparticles of different sizes from solution has been developed by Liang et. al. The produced membranes showed good flexibility, also due to the high mechanical properties of the membrane it withstands high pressure without any damage. The porous structure and hydrophilic properties of the fibrous membranes allowed easier transport [59].

Other types of polymeric membranes were used for the filtration of microorganisms from natural waters. Membrane filters with different functional pore size ranges can be employed with appropriate media to selectively extract target species. The most common filter pore size for microbiological procedures is 0.45 μm [60].

2.5.2 Biomedical Applications

Polymeric nanofibers are beneficial for numerous biomedical applications, such as tissue engineering, drug delivery, protective clothing, wound dressing, etc. These could be made up of blends and copolymers and polymers including biodegradable polymers and nondegradable polymers such as polyacrylonitrile, nylon, polyurethane, polycarbonate, polyethylene oxide, polycaprolactone, polyhydroxy acid, and polyanhydrides.

2.5.2.1 Tissue engineering

The creation of suitable scaffold matrixes that can simulate the biological functions and structure of the natural extracellular matrix (ECM) for the treatment of tissues or organs in the event of any malfunction in the human body is a key challenge in the field of tissue engineering. Human cells can appropriately join and organize themselves around fibers that are smaller in diameter than their own [61].

Tissue engineering is an interdisciplinary field that aims to create scaffolds for tissue and organ regeneration utilizing a variety of synthetic and natural polymer processing techniques [62].

Favorable nanofibrous structures of biodegradable polymers serve as a support for the adhesion, proliferation, and differentiation of diverse cells, and thus have significant potential in tissue regeneration scaffolds [63]. The natural ECM is a complex structure that is customized according to the tissue and organ's specific needs. ECM is mostly made up of nanometer-sized fibrils, but it may also contain other essential elements such as proteoglycans, glycosaminoglycans, and minerals. The goal of tissue engineering research is to copy the ECM in such a way that it provides the same environment [63–65]. Goldberg et al. reported about a variety of nanostructured materials for tissue engineering and drug delivery applications [66].

2.5.2.2 Drug delivery

Topical drug release brings new possibilities for the effective treatment of skin and wound-related issues. The electrospinning technique can be implemented to distribute drugs for the treatment of various types of diseases.

Electrospun fibers are mostly delivered topically, orally, and as implantable systems [67]. Various nonbiodegradable polymers were used for this purpose such as polyurethane-based polymer/drug solutions that were electrospun from dimethylacetamide and dimethylformamide (DMF) [68]. In another study bioabsorbable poly(lactic acid) nanofiber membranes were employed to carry the antibiotic Mefoxin [69].

Suwantong et al. [70] employed pure Asiaticoside and plant crude extract as drug forms. They concluded that the pure drug had a superior release profile than the immersion-assayed extract. They also examined the release profiles of the solvent-cast film, which were significantly lower. Additionally, they reported that the extract-loaded films were nontoxic to the cells at different extraction concentrations. Another study by Taepaiboon et al. shows release characteristics of drug-loaded polyvinyl alcohol fibers which were fabricated by using electrospinning [71].

Bacterial cellulose (BC) mats are used to treat skin conditions such as burns and ulcers. Its pores allow for the exchange of gases between the organism and the surrounding environment. Electrospun BC membranes are used as affinity membranes because of the fibrous biopolymers which help in cell growth [72].

2.5.2.3 Wound dressing

Wound dressings are important for wound healing because they must provide the best possible environment in order to restore skin integrity [73]. Fine biodegradable polymer fibers can be applied to the wounded skin to generate a fibrous matt dressing for the treatment of human skin wounds and burns. This helps in the restoration of normal skin on wounds and prevents the formation of scars [74, 75].

Another group of researchers produced polymeric fibrous patches for wound dressing by using electrospinning which can be able to release amoxicillin an antibiotic drug precisely [76]. Another study employed electrospinning to make poly (lactide-co-glycolide) based nanofibers to see how the morphology and diameter of electrospun nanofibers were affected by the concentration of the solution, orifice diameter, and voltage for wound healing and drug delivery [77].

2.5.3 Applications in Protective Clothing

Soldiers' protective clothing is made exclusively of charcoal. It has several drawbacks, including low water/air permeability, high weight, and so forth. As a result, a light and breathable fabric with good permeability with high reactivity to lethal chemical agents are desirable. Nanofiber-based protective clothing is capable of neutralizing toxic substances. This will improve the individual soldier system's survivability, durability, and combat efficiency in the face of adverse weather, ballistics, and NBC (nuclear, biological, and chemical) warfare. Electrospinning is widely used to produce such kinds of clothing [78]. Lee et al. investigated the layered fabric system with an electrospun polyurethane nanofibrous web to deliver a combination of thermal comfort and good liquid barrier performance for agricultural applications [79].

2.5.4 Applications in Energy Storage

A few methods are used to prepare porous polymeric separators/ membranes for stable and durable battery manufacturing. Some techniques can be followed to prepare polymeric membranes, such as solvent casting, phase separation methods using a non-solvent, extraction of plasticizer, electrospinning, etc. [80]. Construction of such membranes should possess the desired level of high ionic conductivity, and chemical, mechanical, electrical, dielectric, and thermal stability that makes them a prime challenge for the related science and technology of membranes/separators. The typical fibrous characteristic of electrospun nanofibers enables them to absorb electrolyte to a large amount, have low ionic resistance, and shows excellent charge/discharge cycles of a battery in the service life. Electrospun PVDF fiber membrane was utilized as a separator for lithium-ion batteries (LIB) by Liang et al. [80]. Electrospinning was also used to prepare membranes with excellent conductivity, good electrochemical stability, and stable cycle performance. Another group of researchers synthesized electrospun polyaniline nanofibers and discovered that they have substantially greater specific capacitance and rate performance than polyaniline powder [81]. An and Ahn used electrospinning paired with H_2 reduction to create coaxial nanofibers with various topologies to improve electrochemical performance [82]. Electrospinning is perhaps the simplest way to make nanofibers with good porosity. Electrospun carbon nanofibers made from polymer precursors such as polybenzimidazole and PAN have sparked a lot of interest [83]. Yanilmaz et al. recently fabricated electrospun SiO_2/nylon 6,6 nanofiber membranes for use as LIB separators and examined their electrochemical performance. When this nanofibrous membrane was used as a separator, it was shown to be effective. The batteries performed better in terms of cycling and rate than those that used a commercial microporous polyolefin membrane [84].

Electrospun nanofibers can also serve as an electrode material for supercapacitors and in LIB for energy storage applications due to their superior electrochemical performance [75].

2.5.5 Applications as Nanocomposites

Electrospinning has recently been recognized as a feasible method for producing polymeric nanofibers. A range of polymers, mostly in solvent solution and molten form, have been successfully electrospun into ultrafine fibers in recent years. Potential applications for such fibers have been identified, including their usage as reinforcement in nanocomposite formulation [87] nanofibers also play a vital role in the creation of nanocomposites. Nanocomposites may have superior mechanical properties than microfibers of the same material, signifying that they will have better structural properties.

Furthermore, nanofiber-reinforced composites may have several advantages that traditional (middle) composites do not have.

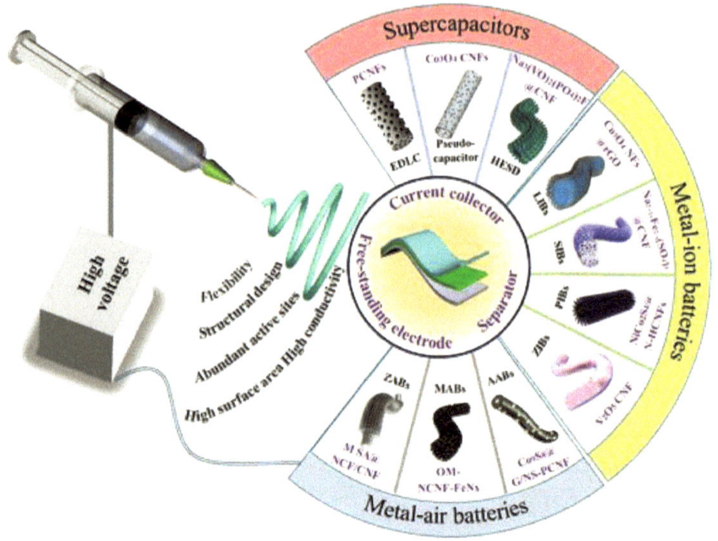

Figure 2.13 Schematic representation of electrospun nanofibers of different materials and their applications in flexible electrochemical energy storage devices [85].

Bognitzki et al. used chemical vapor deposition (CVD) and physical vapor deposition (PVD) coating to create polymer tubes [poly(*p*-xylylene)], composites of polymer and metal (aluminum) nanotubes, and then thermal degradation utilizing PLA [poly(L-lactide)] nanofibers. The tubes' wall thickness ranged from 0.1 to 1 m [86] and polybenzimidazole (PBI) was found as a potential polymer. Furthermore, due to the thermosetting feature resulting from the high stiffness of the polymer chains, researchers anticipate that PBI will give a high carbon yield [87].

2.5.6 Environmental Application

Intensive industrial and urban activities around the world result in severe water and air pollution, causing serious environmental issues as well as negative health repercussions. To effectively remove water and air contaminants and heavy metals and develop self-cleaning surfaces, the cost-effectiveness of nanostructured materials for environmental applications is thought to be of paramount importance. Electrospinning is widely employed for producing nanofibers for environmental applications. Nanofibers with a high surface-to-volume ratio, high permeability, high porosity, and outstanding mechanical qualities are required for environmental applications such as water remediation and air filtering [85, 88].

2.5.7 Applications in Sensors

The properties of nanofibers make them suitable for sensor applications. The high surface-area-to-volume ratio creates a large number of regions and channels, which improves the interaction between the nanofibrous membranes and the determinant and so enhances sensitivity. Furthermore, the use of nanofibrous membranes helps minimize the sensor's energy consumption and overall size, hence broadening its use [89].

Poly(lactic-co-glycolic acid) (PLGA) polymers were studied under various ambient conditions for chemical and biochemical sensors [90]. PVDF (polyvinylidene fluoride) is the most important piezoelectric polymer in terms of technology. Electrospinning has been discovered to be an especially effective technique for synthesizing PVDF nanofibers with exceptional piezoelectric characteristics due to the high fraction of the piezoelectrically active crystalline phase [91]. Nanofibers from polymers with the piezoelectric effect, such as polyvinylidene fluoride/carbon nanotube nanofibrous membranes will render the resulting nanofibrous devices piezoelectric [92].

2.6 Conclusion

There is a growing demand for nanofibers in various fields starting from face masks to wound dressing, controlled drug release, ultrafiltration, sensors, and catalysts. Due to the upgradation of technologies and science, with a better understanding of it, and huge demand in various fields several kinds of nanofibers or nanofibers-based assemblies have been created for a definite purpose. Among all these technologies the electrospinning method for fiber production has been found to be very effective in terms of productivity, quality, variety, and suitability for various applications. Used in combination with various technologies they are creating possibilities in various advanced applications in human life. In the future, more such possibilities in hi-tech usage are expected.

References

1. Khajavi, R. and Damerchely, R. (2007). Effect of polyvinyl alcohol concentration in spinning dope on diameter. *Pak. J. Biol. Sci.*, 10(2), 314–317.
2. Ko, F. K. and Wan, Y. (2014). *Introduction to Nanofiber Materials.* Cambridge University Press, UK.
3. Bhardwaj, N. and Kundu, S. C. (2010). Electrospinning: a fascinating fiber fabrication technique. *Biotechnol. Adv.*, 28(3), 325–347.
4. Zahmatkeshan, M., Adel, M., Bahrami, S., Esmaeili, F., Rezayat, S. M., Saeedi, Y., Mehravi, B., Jameie, S. B., and Ashtari, K. (2019). Polymer-based nanofibers: preparation, fabrication, and applications, in *Handbook of Nanofibres*, Barhoum, A., Bechelany, M., and Makhlouf, A. S. H. (eds.) Springer Nature, Switzerland AG, 215–262.
5. Ondarcuhu, T. and Joachim, C. (1998). Drawing a single nanofibre over hundreds of microns. *Europhys. Lett.*, 42(2), 215.
6. Ma, J., Zhang, Q., Zhang, Y., Zhou, L., Yang, J., and Ni, Z. (2016). A rapid and simple method to draw polyethylene nanofibers with enhanced thermal conductivity. *Appl. Phys. Lett.*, 109(3), 1–5.
7. Nain, A. S., Sitti, M., Jacobson, A., Kowalewski, T., and Amon, C. (2009). Dry spinning-based spinneret based tunable engineered parameters (STEP) technique for controlled and aligned deposition of polymeric nanofibers. *Macromol. Rapid Commun.*, 30(16), 1406–1412.
8. Martin, C. R. (1995). Template synthesis of electronically conductive polymer nanostructures. *Acc. Chem. Res.*, 28(2), 61–68.
9. Alghoraibi, I. and Alomari, S. (2018). Different methods for nanofiber design and fabrication, in *Handbook of Nanofibres*, Barhoum, A., Bechelany, M., and Makhlouf, A. S. H. (eds.) Springer Nature, Switzerland AG, 1–46.
10. Che, G., Lakshmi, B. B., Martin, C. R., Fisher, E. R., and Ruoff, R. S. (1998). Chemical vapor deposition-based synthesis of carbon nanotubes and nanofibers using a template method. *Chem. Mater.*, 10(1), 260–267.
11. Jayaraman, K., Kotaki, M., Zhang, Y., Mo, X., and Ramakrishna, S. (2004). Recent advances in polymer nanofibers. *J. Nanosci. Nanotechnol.*, 4(1–2), 52–65.

12. Ramakrishna, S. (2005). *An Introduction to Electrospinning and Nanofibers*, World Scientific Pub Co Inc, Singapore.
13. He, L., Zuo, Q., Shi, Y., and Xue, W. (2014). Microstructural characteristics and crystallization behaviors of poly (l-lactide) scaffolds by thermally induced phase separation. *J. Appl. Polym. Sci.*, 131(4), 39436 (1–8).
14. Zhang, Z., Hu, J., and Ma, P. X. (2012). Nanofiber-based delivery of bioactive agents and stem cells to bone sites. *Adv. Drug Deliv. Rev.*, 64, 1129–1141.
15. Holzwarth, J. M. and Ma, P. X. (2011). 3D nanofibrous scaffolds for tissue engineering. *J. Mater. Chem.*, 21, 10243–10251.
16. Woo, K. M., Chen, V. J., and Ma, P. X. (2003). Nano-fibrous scaffolding architecture selectively enhances protein adsorption contributing to cell attachment. *J. Biomed. Mater. Res. A.*, 67(2), 531–537.
17. Teo, W. E. and Ramakrishna, S. (2006). A review on electrospinning design and nanofibre assemblies. *Nanotechnology*, 17(14), R89.
18. Whitesides, G. M., Mathias, J. P., and Seto, C. T. (1991). Molecular self-assembly and nanochemistry: a chemical strategy for the synthesis of nanostructures. *Science*, 254(5036), 1312–1319.
19. Zhang, Z., Hu, J., and Ma, P. X. (2012). Nanofiber-based delivery of bioactive agents and stem cells to bone sites. *Adv. Drug Deliv. Rev.*, 64(12), 1129–1141.
20. Kunitake, T. (1992). Synthetic bilayer membranes: molecular design, self-organization, and application. *Angew. Chem. Int. Ed. Engl.*, 31(6), 709–726.
21. Fuhrhop, J. H. and Koning, J. (2007). *Membranes and Molecular Assemblies: The Synkinetic Approach*, The Royal Society of Chemistry, Cambridge, UK.
22. Fuhrhop, J. H. and Wang, T. (2004). Bolaamphiphiles. *Chem. Rev.*, 104(6), 2901–2938.
23. Shimizu, T., Iwaura, R., Masuda, M., Hanada, T., and Yase, K. (2001). Internucleobase-interaction-directed self-assembly of nanofibers from homo- and heteroditopic 1, ω-nucleobase bolaamphiphiles. *J. Am. Chem. Soc.*, 123(25), 5947–5955.
24. Kimizuka, N. (2005). Chapter 13: Soluble amphiphilic nanostructures and potential applications, in *Supramolecular Polymers*, Ciferri, A. (ed.), CRC Press, Boca Raton, 495–522.
25. Fleming, S. and Ulijn, R. V. (2014). Design of nanostructures based on aromatic peptide amphiphiles. *Chem. Soc. Rev.*, 43(23), 8150–8177.

26. Tsutsumi, H. and Mihara, H. (2013). Soft materials based on designed self-assembling peptides: from design to application. *Mol. BioSyst.*, 9(4), 609–617.

27. Chen, C., Gu, Y., Deng, L., Han, S., Sun, X., Chen, Y., Lu, J. R., and Xu, H. (2014). Tuning gelation kinetics and mechanical rigidity of β-hairpin peptide hydrogels via hydrophobic amino acid substitutions. *ACS Appl. Mater. Interfaces*, 6(16), 14360–14368.

28. Pashuck, E. T. and Stupp, S. I. (2010). Direct observation of morphological tranformation from twisted ribbons into helical ribbons. *J. Am. Chem. Soc.*, 132(26), 8819–8821.

29. Lee, M., Cho, B. K., and Zin, W. C. (2001). Supramolecular structures from rod–coil block copolymers. *Chem. Rev.*, 101(12), 3869–3892.

30. Klok, H. A., Langenwalter, J. F., and Lecommandoux, S. (2000). Self-assembly of peptide-based diblock oligomers. *Macromolecules*, 33(21), 7819–7826.

31. Chen, J. T., Thomas, E. L., Ober, C. K. and Mao, G. P. (1996). Self-assembled smectic phases in rod-coil block copolymers. *Science*, 273(5273), 343–346.

32. Hoeben, F. J., Jonkheijm, P., Meijer, E. W., and Schenning, A. P. (2005). About supramolecular assemblies of π-conjugated systems. *Chem. Rev.*, 105(4), 1491–1546.

33. Huang, W. S., Humphrey, B. D., and MacDiarmid, A. G. (1986). Polyaniline, a novel conducting polymer. Morphology and chemistry of its oxidation and reduction in aqueous electrolytes. *J. Am. Chem. Soc.*, 82(8), 2385–2400.

34. Wu, C. G. and Bein, T. (1994). Conducting polyaniline filaments in a mesoporous channel host. *Science*, 264(5166), 1757–1759.

35. Huang, J., Virji, S., Weiller, B. H., and Kaner, R. B. (2004). Nanostructured polyaniline sensors. *Chem. Eur. J.*, 10(6), 1314–1319.

36. Sun, Z., Zussman, E., Yarin, A. L., Wendorff, J. H., and Greiner, A. (2003). Advanced Materials (Weinheim, Germany), 15, 1929–1932.

37. Yu, J. H., Fridrikh, S. V., and Rutledge, G. C. (2004). *Adv. Mater.*, 16(17), 1562–1566.

38. Li, D., Babel, A., Jenekhe, S., and Xia, Y. (2004). Nanofibers of conjugated polymers prepared by electrospinning with a two-capillary spinneret, *Adv. Mater.*, 16 (22), 2062–2066.

39. Subrahmanya, T. M., Arshad, A. B., Lin, P. T., Widakdo, J., Makari, H. K., Austria, H. F. M., Hu, C. C., Lai, J. Y., and Hung, W. S. (2021). A review

of recent progress in polymeric electrospun nanofiber membranes in addressing safe water global issues. *RSC Adv.*, 11(16), 9638–9663.

40. Yarin, A. L., Pourdeyhimi, B., and Ramakrishna, S. (2014). *Fundamentals and Applications of Micro- and Nanofibers*. Cambridge University Press.

41. Vonnegut, B. and Neubauer, R. L. (1952). Production of monodisperse liquid particles by electrical atomization, *J. Colloid Sci.*, 7(6), 616–622.

42. Rayleigh, L. (1882). XX. On the equilibrium of liquid conducting masses charged with electricity. *Lond. Edinb. Dublin Philos. Mag. J. Sci.*, 14(87), 184–186.

43. Taylor, G. I. (1969). Electrically driven jets. *Proc. R. Soc. Lond. A.*, 313(1515), 453–475.

44. Reneker, D. H., Yarin, A. L., Fong, H., and Koombhongse, S. (2000). Bending instability of electrically charged liquid jets of polymer solutions in electrospinning. *J. Appl. Phys.*, 87(9), 4531–4547.

45. Yarin, A. L., Pourdeyhimi, B., and Ramakrishna, S. (2014). Chapter 5: Electrospinning of micro- and nanofiber, In *Fundamentals and Applications of Micro- and Nanofibers*, p. 179. Cambridge University Press.

46. Taylor, G. I. (1964). Disintegration of water drops in an electric field. *Proc. R. Soc. Lond. A.*, 280(1382), 383–397.

47. Brown, P. and Stevens, K. (eds.) (2007). *Nanofibers and Nanotechnology in Textiles*. Elsevier.

48. Reneker, D. H., Yarin, A. L., Fong, H., and Koombhongse, S. (2000). Bending instability of electrically charged liquid jets of polymer solutions in electrospinning. *J. Appl. Phys.*, 87(9), 4531–4547.

49. James, S. J. (1966). *Mathematical Theory of Electricity and Magnetism*. Cambridge University Press, New York.

50. Reneker, D. H., and Fong, H. (2006). *Polymeric Nanofibers: Introduction*. ACS Symposium Series, No. 918, Oxford University Press.

51. Gopal, R., Kaur, S., Ma, Z., Chan, C., Ramakrishna, S., and Matsuura, T. (2006). Electrospun nanofibrous filtration membrane, *J. Membr. Sci.*, 281(1–2), 581–586.

52. Ahmed, F. E., Lalia, B. S., and Hashaikeh, R. (2015). A review on electrospinning for membrane fabrication: challenges and applications. *Desalination*, 356, 15–30.

53. Tsai, P. P., Schreuder-Gibson, H., and Gibson, P. (2002). Different electrostatic methods for making electret filters. *J. Electrostat.*, 54(3–4), 333–341.

54. Ko, F. K. and Wan, Y. (2014). *Introduction to Nanofiber Materials*. Cambridge University Press.

55. Sundarrajan, S., Tan, K. L., Lim, S. H., and Ramakrishna, S. (2014). Electrospun nanofibers for air filtration applications. *Procedia Eng.*, 75, 159–163.

56. Lala, N. L., Ramaseshan, R., Bojun, L., Sundarrajan, S., Barhate, R. S., Ying-jun, L., and Ramakrishna, S. (2007). Fabrication of nanofibers with antimicrobial functionality used as filters: protection against bacterial contaminants. *Biotechnol. Bioeng.*, 97(6), 1357–1365.

57. Gopal, R., Kaur, S., Feng, C. Y., Chan, C., Ramakrishna, S., Tabe, S., and Matsuura, T. (2007). Electrospun nanofibrous polysulfone membranes as pre-filters: particulate removal. *J. Membr. Sci.*, 289(1–2), 210–219.

58. Huang, Z. M., Zhang, Y. Z., Kotaki, M., and Ramakrishna, S. (2003). A review on polymer nanofibers by electrospinning and their applications in nanocomposites. *Compos. Sci. Technol.*, 63(15), 2223–2253.

59. Liang, H. W., Wang, L., Chen, P. Y., Lin, H. T., Chen, L. F., He, D., and Yu, S. H. (2010). Carbonaceous nanofiber membranes for selective filtration and separation of nanoparticles. *Adv. Mater.*, 22(42), 4691–4695.

60. Kator, H. and Rhodes, M. (2003). Detection, enumeration and identification of environmental microorganisms of public health significance, in *The Handbook of Water and Wastewater Microbiology*, Mara, D. and Horan, N. (eds.), Elsevier, 113–126.

61. O'brien, F. J. (2011). Biomaterials and scaffolds for tissue engineering. *Mater Today*, 14(3), 88–95.

62. Barnes, C. P., Sell, S. A., Boland, E. D., Simpson, D. G., and Bowlin, G. L. (2007). Nanofiber technology: designing the next generation of tissue engineering scaffolds. *Adv. Drug Deliv. Rev.*, 59(14), 1413–1433.

63. Mathuriya, A. S., Yakhmi, J. V., Martínez, L. M. T., Kharissova, O. V., and Kharisov, B. I. (2017). Polyhydroxyalkanoates: biodegradable plastics and their applications, in *Handbook of Ecomaterials*, Martínez, L. M. T., Kharissova, O. V., and Kharisov, B. I. (eds.), Springer, Cham, 1–29.

63. Teo, W. E., He, W. and Ramakrishna, S. (2006). Electrospun scaffold tailored for tissue-specific extracellular matrix. *Biotechnol. J.*, 1(9), 918–929.

64. Cui, W., Zhou, Y. and Chang, J. (2010). Electrospun nanofibrous materials for tissue engineering and drug delivery. *Sci. Technol. Adv. Mater.*, 11(1), 014108.

65. Unnithan, A. R., Arathyram, R. S., and Kim, C. S. (2015). Electrospinning

of polymers for tissue engineering, in *Nanotechnology Applications for Tissue Engineering*, Thomas, S., Grohens, Y., and Ninan, N. (eds.), William Andrew Publishing, Elsevier, 45–55.

66. Goldberg, M., Langer, R., and Jia, X. (2007). Nanostructured materials for applications in drug delivery and tissue engineering. *J. Biomater. Sci. Polym. Ed.*, 18(3), 241–268.
67. Bhattarai, R. S., Bachu, R. D., Boddu, S. H., and Bhaduri, S. (2019). Biomedical applications of electrospun nanofibers: drug and nanoparticle delivery. *Pharmaceutics*, 11(1), 5.
68. Verreck, G., Chun, I., Rosenblatt, J., Peeters, J., Van Dijck, A., Mensch, J., Noppe, M., and Brewster, M. E. (2003). Incorporation of drugs in an amorphous state into electrospun nanofibers composed of a water-insoluble, nonbiodegradable polymer. *J. Control. Release*, 92(3), 349–360.
69. Zussman, E., Yarin, A., and Weihs, D. (2002). A micro-aerodynamic decelerator based on permeable surfaces of nanofiber mats. *Exp. Fluids*, 33(2), 315–320.
70. Suwantong, O., Ruktanonchai, U., and Supaphol, P. (2008). Electrospun cellulose acetate fiber mats containing asiaticoside or *Centella asiatica* crude extract and the release characteristics of asiaticoside, *Polymer*, 49(19), 4239–4247.
71. Taepaiboon, P., Rungsardthong, U., and Supaphol, P. (2006). Drug-loaded electrospun mats of poly (vinyl alcohol) fibres and their release characteristics of four model drugs, *Nanotechnology*, 17(9), 2317.
72. Ma, Z., Kotaki, M., and Ramakrishna, S. (2005). Electrospun cellulose nanofiber as affinity membrane. *J. Membr. Sci.*, 265(1–2), 115–123.
73. Sandri, G., Rossi, S., Bonferoni, M. C., Caramella, C., and Ferrari, F. (2020). Electrospinning technologies in wound dressing applications, in *Therapeutic Dressings and Wound Healing Applications*, Boateng, J. (ed.), Advances in Pharmaceutical Technology, Wiley, 315–336.
74. Rho, K. S., Jeong, L., Lee, G., Seo, B. M., Park, Y. J., Hong, S. D., Roh, S., Cho, J. J., Park, W. H., and Min, B. M. (2006). Electrospinning of collagen nanofibers: effects on the behavior of normal human keratinocytes and early-stage wound healing. *Biomaterials*, 27(8), 1452–1461.
75. Asmatulu, R. and Khan, W. S. (2018). *Synthesis and Applications of Electrospun Nanofibers*, Elsevier.
76. Sofokleous, P., Stride, E., and Edirisinghe, M. (2013). Preparation, characterization, and release of amoxicillin from electrospun fibrous wound dressing patches, *Pharm. Res.*, 30(7), 1926–1938.
77. Katti, D. S., Robinson, K. W., Ko, F. K., and Laurencin, C. T. (2004).

Bioresorbable nanofiber-based systems for wound healing and drug delivery: optimization of fabrication parameters. *J. Biomed. Mater. Res. B Appl. Biomater.*, 70(2), 286–296.

78. Gorji, M., Bagherzadeh, R., and Fashandi, H. (2017). Electrospun nanofibers in protective clothing, in *Electrospun Nanofibers*, Afshari, M. (ed.), Woodhead Publishing Series in Textiles, Elsevier, 571–598.

79. Lee, S. and Obendorf, S. K. (2007). Use of electrospun nanofiber web for protective textile materials as barriers to liquid penetration. *Text. Res. J.*, 77(9), 696–702.

80. Liang, Y., Cheng, S., Zhao, J., Zhang, C., Sun, S., Zhou, N., Qiu, Y., and Zhang, X. (2013). Heat treatment of electrospun polyvinylidene fluoride fibrous membrane separators for rechargeable lithium-ion batteries, *J. Power Sources*, 240, 204–211.

81. Chaudhari, S., Sharma, Y., Archana, P. S., Jose, R., Ramakrishna, S., Mhaisalkar, S., and Srinivasan, M. (2013). Electrospun polyaniline nanofibers web electrodes for supercapacitors. *J. Appl. Polym. Sci.*, 129(4), 1660–1668.

82. An, G. H. and Ahn, H. J. (2013). Activated porous carbon nanofibers using Sn segregation for high-performance electrochemical capacitors. *Carbon*, 65, 87–96.

83. Shi, X., Zhou, W., Ma, D., Ma, Q., Bridges, D., Ma, Y., and Hu, A. (2016). Electrospinning of nanofibers and their applications for energy devices. *Nanomaterials*, 6(7), 129.

84. Yanilmaz, M., Dirican, M., and Zhang, X. (2014). Evaluation of electrospun SiO_2/nylon 6,6 nanofiber membranes as a thermally-stable separator for lithium-ion batteries, *Electrochimica Acta*, 133, 501–508.

85. Yan, Y., Liu, X., Yan, J., Guan, C., and Wang, J. (2021). Electrospun nanofibers for new generation flexible energy storage. *Energy Environ. Mater.*, 4(4), 502–521.

86. Bognitzki, M., Hou, H., Ishaque, M., Frese, T., Hellwig, M., Schwarte, C., Schaper, A., Wendorff, J. H., and Greiner, A. (2000). Polymer, metal, and hybrid nano-and mesotubes by coating degradable polymer template fibers (TUFT process), *Adv. Mater.*, 12(9), 637–640.

87. Kim, C., Yang, K. S., Kojima, M., Yoshida, K., Kim, Y. J., Kim, Y. A., and Endo, M. (2006). Fabrication of electrospinning-derived carbon nanofiber webs for the anode material of lithium-ion secondary batteries, *Adv. Funct. Mater.*, 16(18), 2393–2397.

88. Christou, C. N. and Krasia-Christoforou, T. (2020). Nanostructured

electrospun fibers in environmental applications, in *Advances in Nanostructured Materials and Nanopatterning Technologies*, Guarino, V., Focarete, M. L., and Pisignano, D. (eds.), Elsevier, 203–241.

89. Li, Y., Abedalwafa, M. A., Tang, L. and Wang, L. (2019). Electrospun nanofibers for sensors, in *Electrospinning: Nanofabrication and Applications*, Ding, B., Wang, X., and Yu, J. (eds.), William Andrew Publishing, Elsevier, 571–601.
90. Kwoun, S. J., Lec, R. M., Han, B., and Ko, F. K. (2000). A novel polymer nanofiber interface for chemical sensor applications, in *Proceedings of the 2000 IEEE/EIA International Frequency Control Symposium and Exhibition* (Cat. No. 00CH37052), IEEE, 52–57.
91. Azzaz, C. M., Mattoso, L. H., Demarquette, N. R., and Zednik, R. J. (2021). Polyvinylidene fluoride nanofibers obtained by electrospinning and blowspinning: electrospinning enhances the piezoelectric β-phase–myth or reality? *J. Appl. Polym. Sci.*, 138(10), 49959.
92. Wu, C. M., Chou, M. H., and Zeng, W. Y. (2018). Piezoelectric response of aligned electrospun polyvinylidene fluoride/carbon nanotube nanofibrous membranes, *Nanomaterials*, 8(6), 420.

Chapter 3

Nanoyarns: Recent Advancements in Production Techniques, Applications, and Future Prospects

Çağlar Sivri
Bahcesehir University, Faculty of Engineering and Natural Sciences, Ciragan Cd. Yali Sk. No. 1, 34349, Besiktas, Istanbul, Turkey
caglar.sivri@eng.bau.edu.tr

3.1 Introduction

Nanofibers have experienced a dramatic change since their invention in terms of fiber morphologies, production techniques, functions added, application areas, and commercial uses. With their high surface-area-to-volume ratio owing to lower fiber diameters, light-weightiness because of ultralow specific gravity, controllable and interconnected pore structures, and versatility in production, nanofibers have proven their success in medical, filtration, energy production, and storage areas [1]. Despite all these superior properties, nanofibers are not yet durable and strong enough for high-performance applications and repeated uses [2]. Therefore,

Handbook of Nanofibers and Nanocomposites: Characteristics, Synthesis, and Applications in Textiles
Edited by Mohd Yusuf and Aminoddin Haji

ISBN 978-981-4968-77-5 (Hardcover), 978-1-003-43274-6 (eBook)
www.jennystanford.com

their production in nanoyarn form has been an important and extensively studied the issue in recent years. This chapter provides a comprehensive overview regarding present production nanoyarn forming techniques; not only the electrospinning technique but also other manufacturing possibilities such as solution (air) blow spinning and centrifugal spinning have been thoroughly examined. Lastly, the prospects and insights into nanoyarn production were also discussed.

3.2 Nanofibers' Timeline and Yarn Production

It has been more than four centuries since their first introduction, nanofibers experienced a dramatic change in terms of production techniques, morphologies, and applications [3]. Novel nanofiber techniques such as solution blowing and centrifugal spinning have been introduced beneath the electrospinning technique and different techniques have advantages over each other. Electrospinning is still the most common method as it is user-friendly and easy to use, easy to set up, highly versatile, and adjustable in design [4]. This reputation can also be attributed to the extensive scientific-purpose usage of the method and there is an enormous number of publications available on this topic. In the very first years of electrospinning, researchers dealt with more physics related to processes such as charges and their control with electricity, surface tension, and its effect on droplet formation [5–8]. Electrospun nanofibers were manufactured for application purposes for the first time in 1938 to filter substances [9]. Later, the nanofibers area faced incremental changes for a long time until the last 20 years when the dramatic changes occurred in a short period. For example, the single function provided by one type of nanofiber couldn't satisfy the basic requirements, therefore, the production of bicomponent nanofibers has become an area of interest. Researchers developed various types of multifunctional bicomponent nanofibers. For example, they have been produced with a self-crimping function in one application [10]. In another application, nanofibers bending with differential shrinking were developed [11]. Few more functions such as controlled substance release, and even, three-component nanofibers were developed further to add more functionality [12, 13]. Researchers also investigated possibilities of 3-dimensional (3D) nanofiber production, especially for tissue engineering applications. To this aim, 3D nanofibers with novel architectures were developed either by adding salt particles, polyelectrolytes, or organic acids to the solution, or using special apparatus such as a water-based vortex-supported setup and via spherical concave collector [14–18]. Last but not least, bio-inspired nanofibers having bubble-like and/or honeycomb architecture or spider web configuration were developed via tailored electrospinning apparatus [19, 20]. On the other hand, centrifugal nanofiber spinning, where centrifugal forces are used instead of electric charges, is respectively a new method for nanofiber production comparison offering new possibilities to work with a different types of polymers and materials that can not be processed by the electrospinning [21]. Even though it was first used to produce micrometer-scale glass fibers in the 1950s [22], it was the first time used for the production of nanofibers commercially by FibeRio Technology Corporation [23]. To get the most out of the advantages of the centrifugal spinning method and electrospinning together, researchers have developed a hybrid centrifugal electrospinning method and apparatus and the results are promising in terms of versatility and application [24, 25]. Solution blowing in other words air-blowing is a new technique for nanofiber production that is becoming widely utilized by the academy and industry in recent years. Solution blow spinning has gained remarkable interest in a short time in the last years; thanks to its speed for nanofiber production making it faster than electrospinning and centrifugal spinning [26]. This technique will be discussed in more detail in later parts of this chapter, especially in relevance to nanofiber yarn production. There are other techniques available for nanofiber production such as melt blowing, wet spinning, template synthesis, phase separation, et cetera, but this chapter has been limited to electrospinning, centrifugal spinning, and solution blow spinning as they are the most scalable and commercial sounding techniques in our day. Spinning from nanofibers dates back to the first half of the 20th century when Formhals issued the first patent even though twisting was not carried out within the spinning process [27]. Adding a twist to the nanofiber bundles and transforming them into a real yarn form have been realized long years after Formhals' work. Twisted nanofibers were either produced from nanofiber layers by cutting them into short tows and twisting these tows using an electric spinner or directly getting twisted yarns using rings integrated into the electrospinning setup [28, 29]. Some nanofiber yarn production work has been reported based on the combination and/or variation of these two techniques. This chapter focuses on and relates to the production techniques of twisted nanofiber yarns.

3.2.1 Properties of Nanoyarns

Nanoyarns hold promises for many different applications with their higher tenacity, durability, and uniformity as a reinforced version of nanofibers. Nanoyarn properties are mainly affected by process parameters as relevant research has shown. Within this context, Sanatgar et al. have developed polyamide 66 nanofiber yarns to obtain a durable material and found that the combination of formic acid and chloroform as a solvent in low concentration polymer solution improves structural and mechanical properties of the PA66 nanofiber yarns [30]. Abbasipour and Khajavi informed that the number of fibers in the unit area cross-section for nanoyarns is very high compared to conventional yarns for constant linear densities. Therefore, the flexibility and handling of nanoyarns are very high [31]. Jin et al. have manufactured nanoyarns from poly(sulfone amide) fibers and found that the wicking ability of nanoyarns increased with increasing applied voltage, thus nanoyarns have provided better water absorption performance [32]. Moghbelnejad et al. have compared tensile properties of two-layer and three-layer composite nanofibrous yarns from polyamide 6/poly(L-lactic acid and a study has shown that two-layer nanofiber yarn has performed higher tensile strength than three-layer nanofiber yarn did [33].

3.3 Nanoyarn Production Techniques

The most common production technique in nanoyarn production is electrospinning as it is a well-known and widely accepted method for manufacturing nanofibers. Centrifugal spinning and air-blowing also hold promise for nanoyarn production, but there is still time needed for these techniques to become widespread. Essential considerations and details of these techniques for nanoyarn production will be discussed in subsequent chapters.

3.3.1 Nanoyarns via Electrospinning

It was reported by Göktepe and Mülayim that the production of nanofibers has begun in the 1930s [34]. Since then, several different electrospinning setups were modified to manufacture nanoyarns in various morphologies and architectures. Since this chapter focuses on current techniques, only studies that took place in the 21st century are included in this section. The up-to-date methods and properties were presented in Table 3.1 briefly.

In 2005, Smit et al. reported a technique where electrospun nanofibers were collected on a water reservoir and drawn with the aid of a glass rod at the same time forming a nanoyarn at last (Table 3.1A). The higher solution concentration leads to nanofiber with an average diameter of 1 mm, while the lower concentration leads to nanofiber with an average diameter of 294 nm. SEM analysis has shown that nanofibers in produced yarns from poly(vinyl acetate), poly(vinylidene difluoride), and polyacrylonitrile are aligned [35]. Wang et al. manufactured continuous polymer nanofiber yarns from poly(3-hydroxybutyrate-co-3-hydroxy-valerate) (PHBV), polyacrylonitrile (PAN), poly(L-lactic acid) (PLLA) and poly(m-phenylene isophthalamide) (PMIA) polymers using the self-bundling electrospinning method in 2008 (Table 3.1B). A grounded needle tip was used to provide the self-bundling for polymer nanofibers at the beginning of the electrospinning process. The average diameter of the nanofiber yarn from PAN fiber was found as 1.5 mm. SEM analysis revealed that self-bundled nanoyarns comprised ideally aligned nanofibers [36]. Bazbouz and Stylios introduced a new electromechanical mechanism for spinning core electrospun nanoyarn in 2009 (Table 3.1C). Researchers determined twist rates at 500 to 750 revolutions per minute at core filament and 1.5 cm/s take-up speed as ideal spinning parameters. SEM observations have shown that manufactured nanofibers were wrapped helically around the core filament and finally a core electrospun nanoyarn was obtained [37].

Table 3.1 Nanoyarns via electrospinning

Code	Setup	Yarn Images	Principle/ Yarn Properties in Brief	Ref.
A	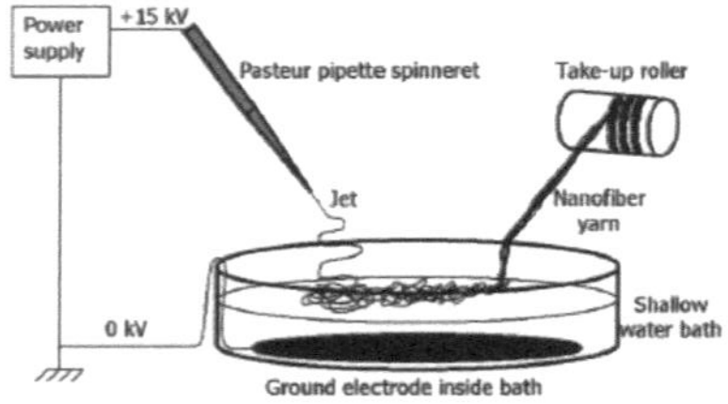	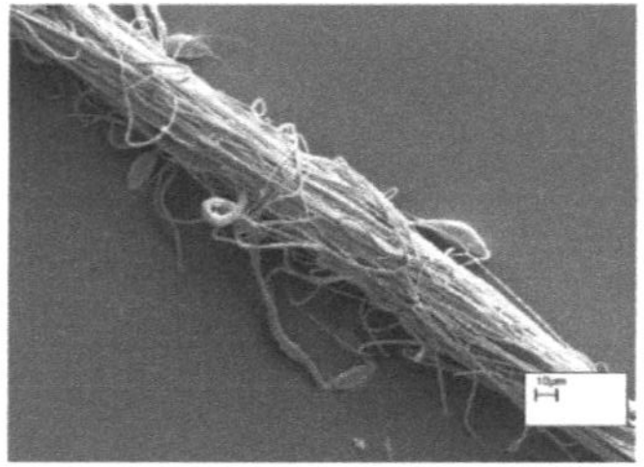	Nanofibers collected on a water reservoir and drawn via glass rod. Nanofibers aligned in yarn.	[35]
B	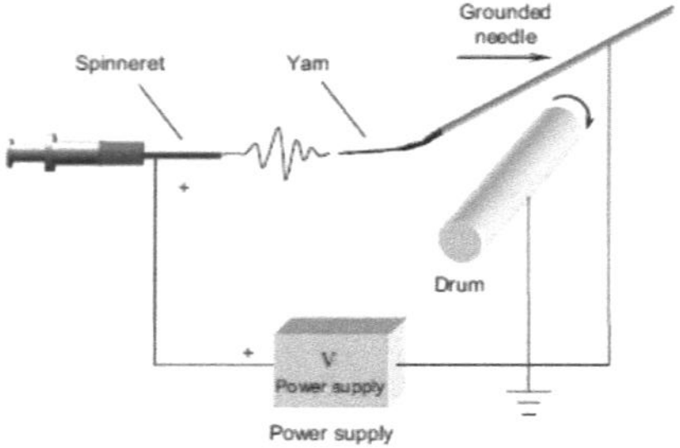	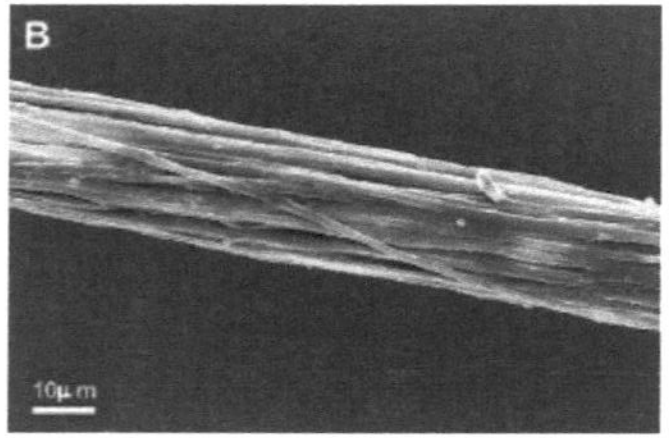	Grounded needle tip for self-bundling. Ideal alignment of nanofibers in yarn.	[36]

Code	Setup	Yarn Images	Principle/ Yarn Properties in Brief	Ref.
C	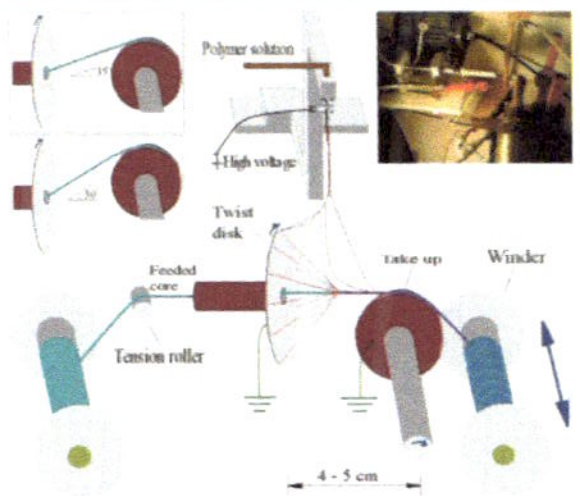	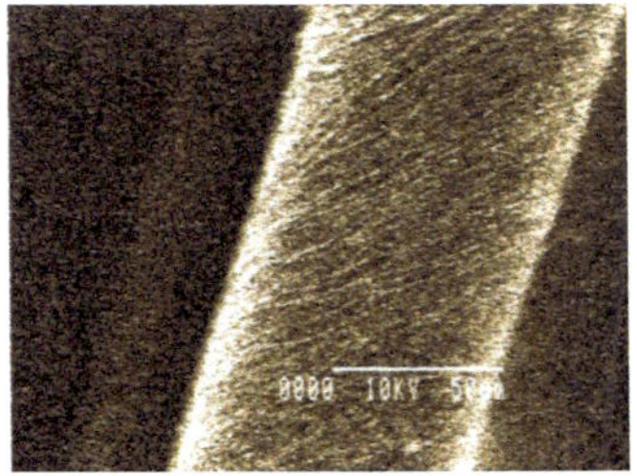	Electromechanical mechanism for spinning. Twist rates at 500 to 750 rpm and 1.5 cm/s take-up speed.	[37]
D	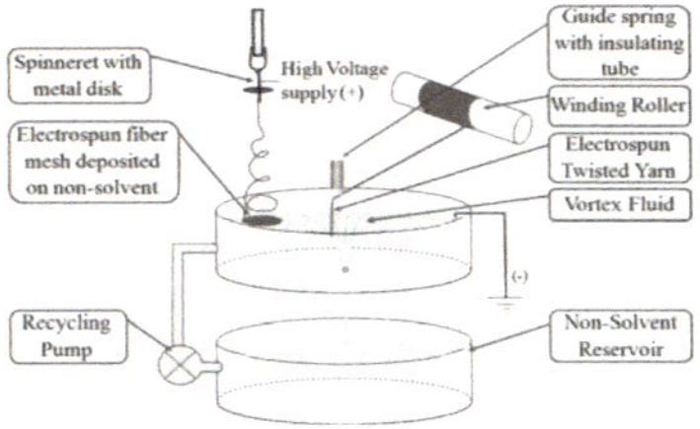	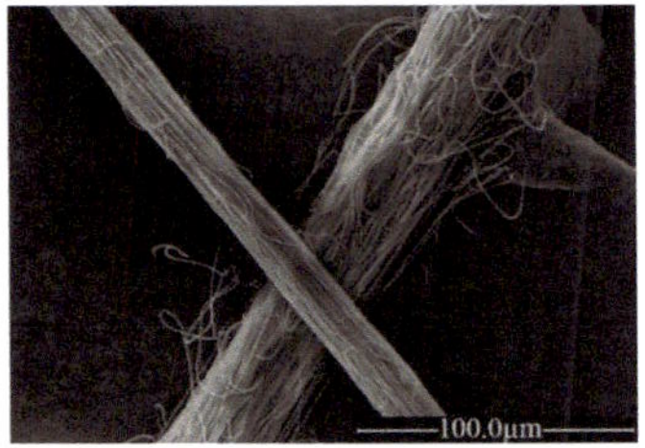	Liquid vortex based spinning for twisted nanofiber yarn. Suitable for weaving and knitting processes	[38]
E	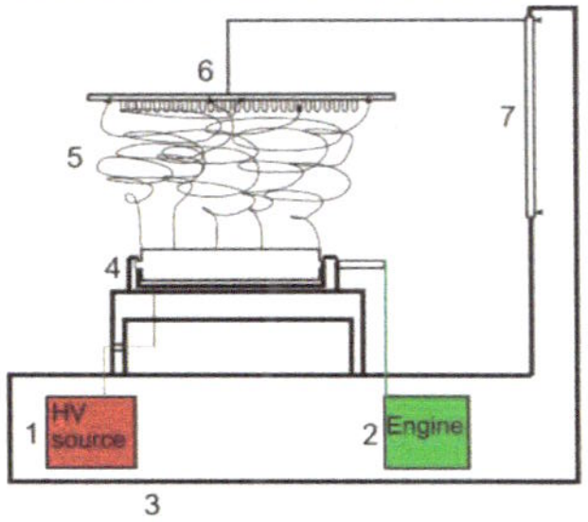	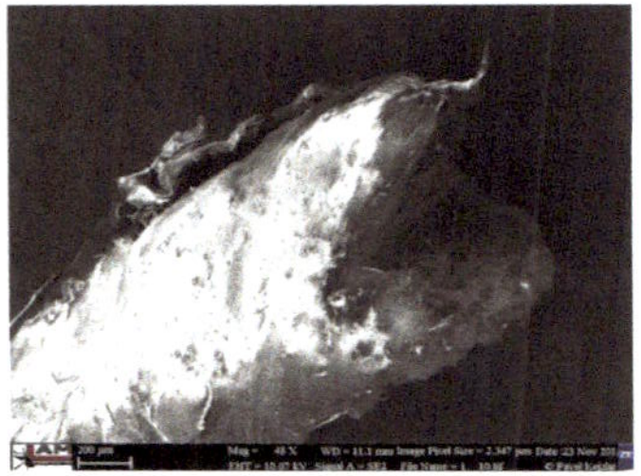	Saw-like collector. Strength between 272,52 and 654,41 cN. Nanoyarn diameter between 0.7 to 1.2 mm	[39]

(Continued)

Table 3.1 *(Continued)*

Code	Setup	Yarn Images	Principle/ Yarn Properties in Brief	Ref.
F	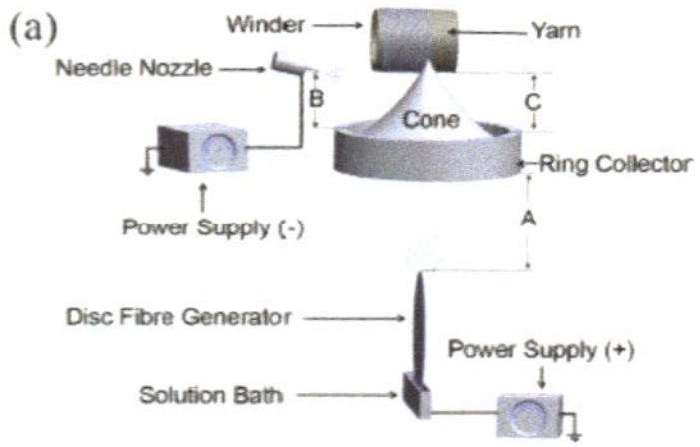	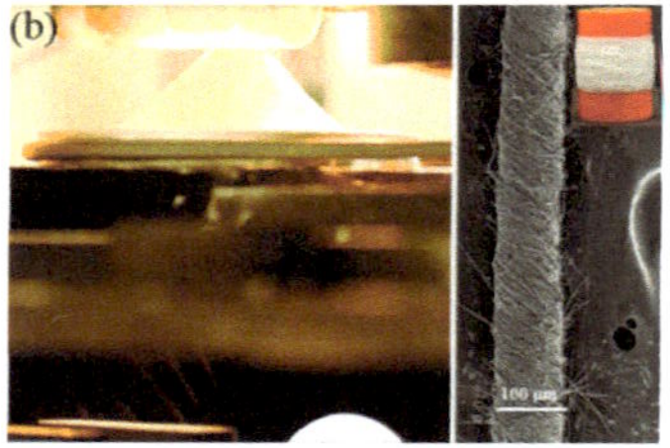	Hybrid electrospinning. Rotating intermediate ring collector. Yarn diameter in the range of 52 μm to 206 μm	[40]
G	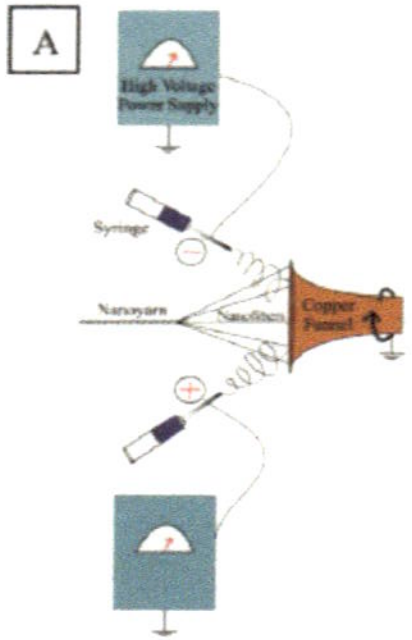	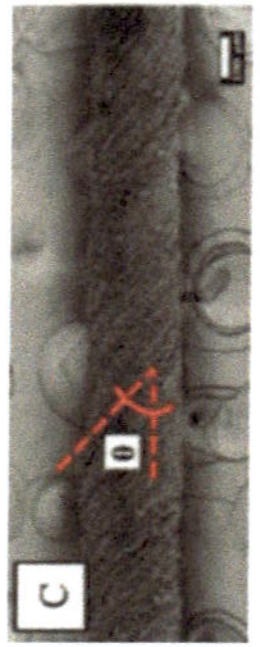	Modified electrospinning setup, alignment using rotating funnel collector. Nanofiber yarns' porosity was studied for the first time	[41]

Code	Setup	Yarn Images	Principle/ Yarn Properties in Brief	Ref.
H	N/A	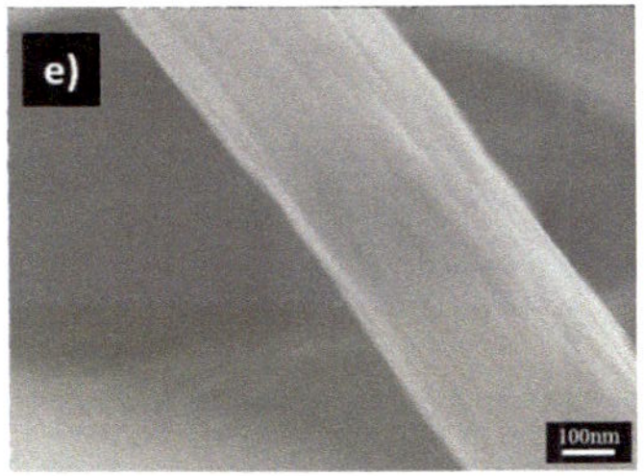	Electrospinning setup consisting of drum collector. Graphene addition improved mechanical properties	[42]
I	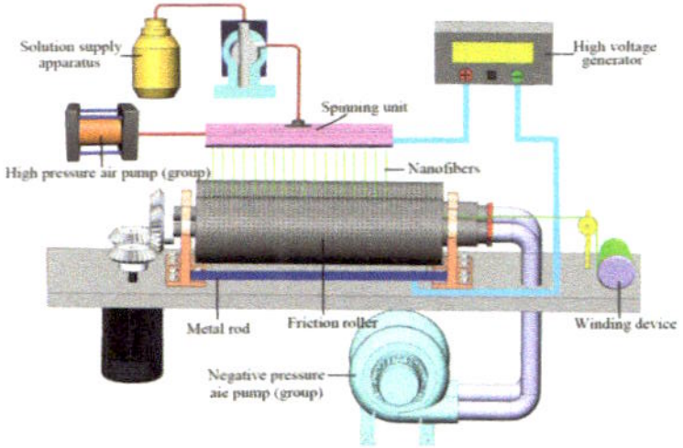	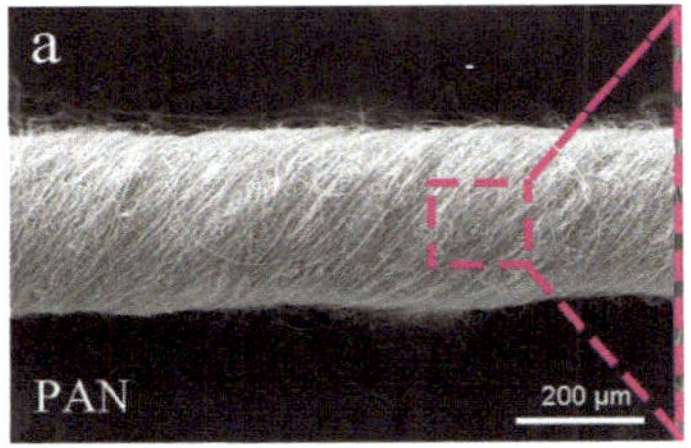	Stepped airflow electrospinning and friction twisting. Excellent alignment and mechanical properties.	[43]
J	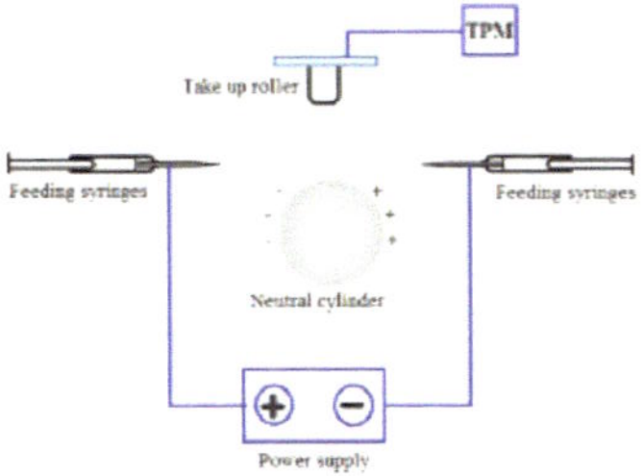	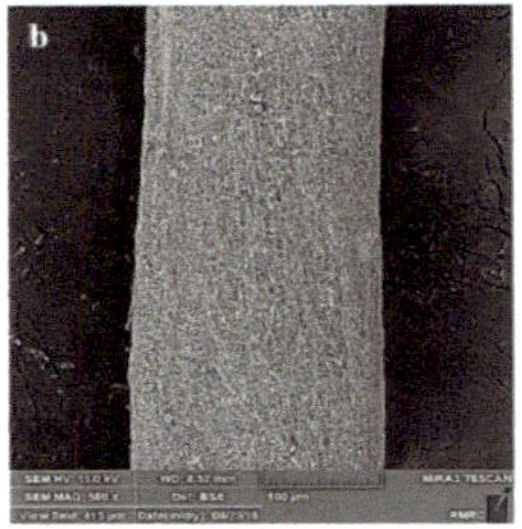	Glass nanoyarn with 2000 TPM tensile strength. Silver nanoparticles for finer fibers. 136 nm fiber diameter.	[44]

(*Continued*)

Table 3.1 *(Continued)*

Code	Setup	Yarn Images	Principle/ Yarn Properties in Brief	Ref.
K	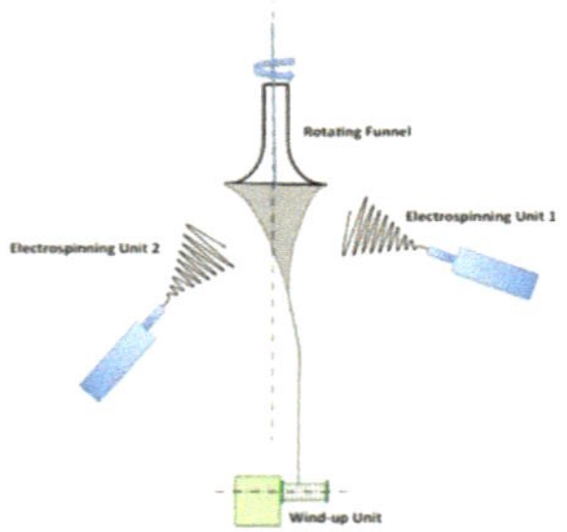	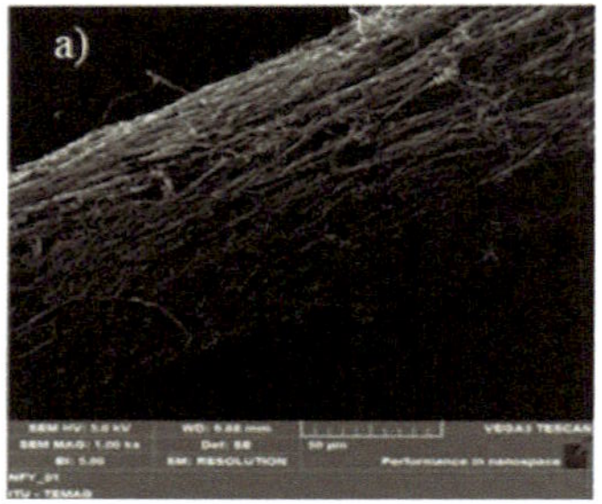	Conventional yarn for drawing. PAN based carbon nanofibers with ≅ 160 nm–300 nm after carbonization.	[45]
L	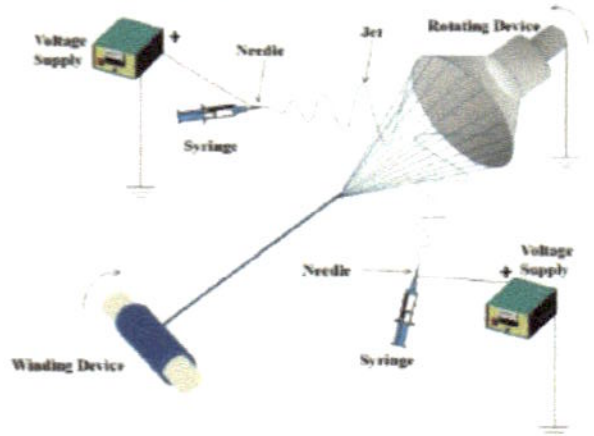	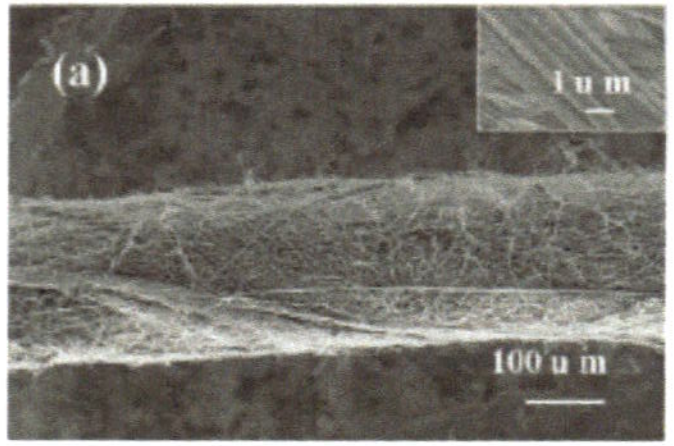	Dynamic rotating electrospinning. Higher rotation speed yields higher number of twist of nanoyarns	[46]
M	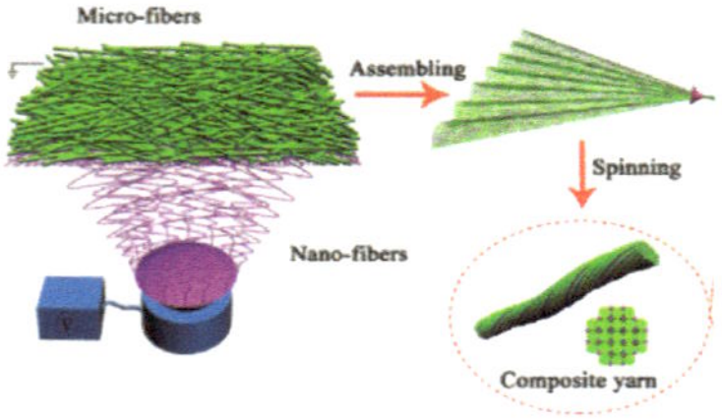	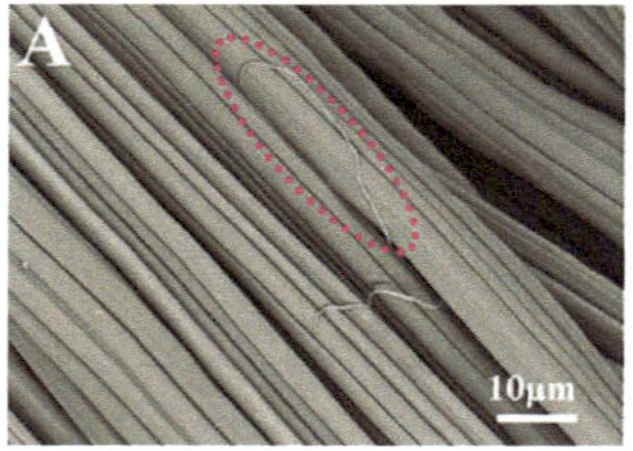	Hybrid strategy for composite micro-/nanoyarn. Durable for preparation, daily use and washing processes.	[47]

Code	Setup	Yarn Images	Principle/ Yarn Properties in Brief	Ref.
N	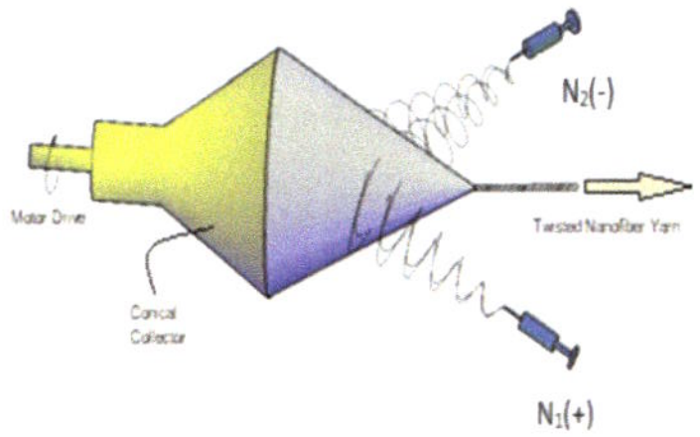	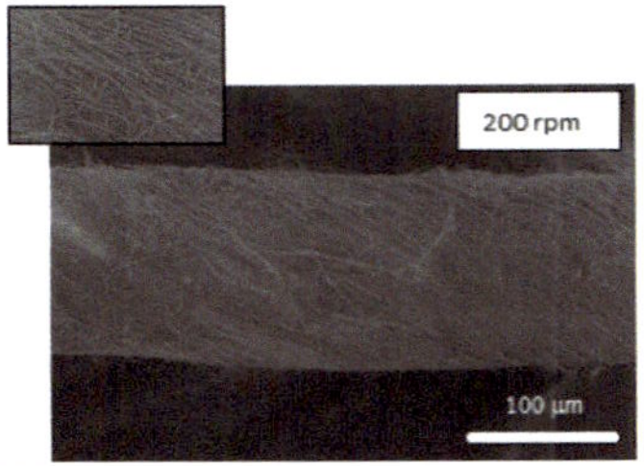	Conical collector with 400 and 500 rpm. Potential system for nanoyarn. The first manually woven fabric from nanoyarn.	[48]
O	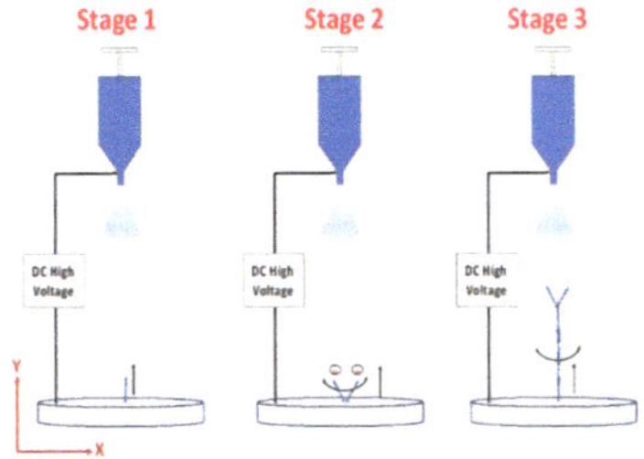	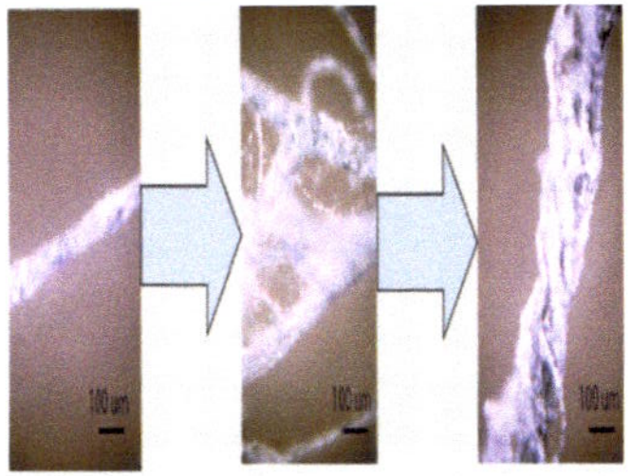	Twisting triggered by current electrospinning. The longer process time, the longer and uniform nano-yarns. Energy-efficient method.	[49]
P	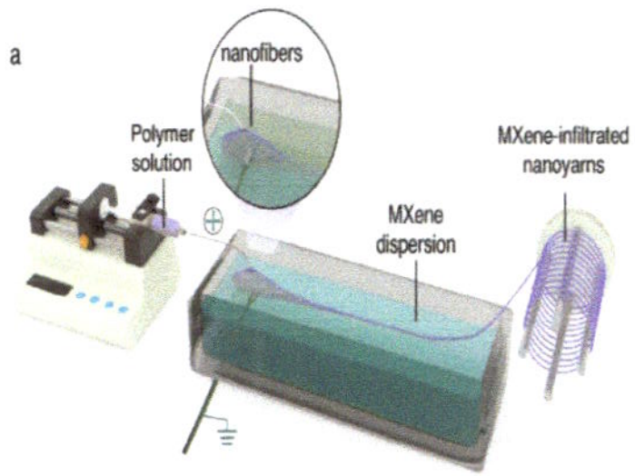	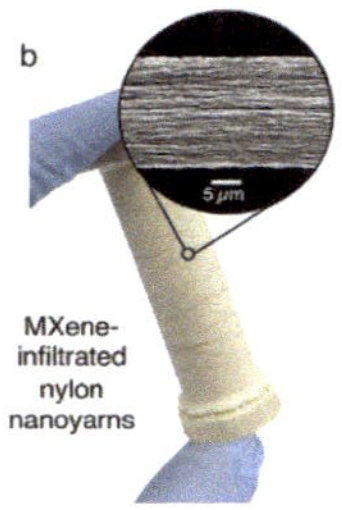	Bath electrospinning. Stretchable nanoyarn up to 43%. Potential for stretchable electronic and body monitoring.	[50]

Yousefzadeh et al. used a dynamic liquid system in the form of a vortex to fabricate continuous twisted nanofiber yarn in 2011 (Table 3.1D). The method enabled continuous production, and simultaneous twisting, and the yarn length is only limited by the amount of solution. Continuous twisted yarn from well-aligned nanofibers was obtained at the end which is promising for weaving and knitting processes [38]. In 2013, Chvojka et al. used a special saw-like collector to align nanofibers in accordance with the electric field and twisted in a separate process twisting device run with a rotation speed of around 750 rpm to form nanoyarns with a length of about 50 cm (Table 3.1E). The strength of individual yarn pieces was measured using a VIBRODYN 400 dynamometer and strength values ranged between 272,52 and 654,41 cN. The average diameter value of nanofibers in obtained yarn was measured as 441.61 nm and the diameter of nanoyarn varied from 0.7 to 1.2 mm [39]. Shuakat and Lin used a hybrid nanofiber yarn electrospinning technique combining needle and needleless electrospinning in 2015 (Table 3.1F). Nanofibers were collected using a rotating intermediate ring collector, drawn, and twisted in nanoyarn form. The system's throughput was 240 m/h as the production rate and twist level were measured up to 4,766 twists per meter. Nanofiber diameter was found in the range of 541 nm to 1.6 μm and yarn diameter in the range of 52 μm to 206 μm [40]. Levitt et al. investigated the nanoyarn preparation from PAN, polyvinylidene fluoride trifluoroethylene (PVDF-TrFe), and polycaprolactone (PCL) polymers using a modified electrospinning setup in 2016 (Table 3.1G). Twisted nanofiber yarn was aligned using a rotating funnel collector. Of all nanofiber yarns investigated, PCL nanofiber yarns showed the highest twist angle, strain-to-failure, and lowest porosity [41]. Mehrpouya et al. produced nanostructured electrospun hybrid graphene/polyacrylonitrile yarns using an electrospinning setup consisting of a drum collector in 2017 (Table 3.1H). Liquid crystal graphene oxide (LCGO) addition has remarkably improved the mechanical properties of PAN/ LCGO composite nanofiber yarn in comparison to PAN-only nanofiber yarn. SEM investigations revealed that nanofibers were placed in yarn structure uniformly for the most part with a helix angle of ~35° to the yarn axis [42]. In 2018, Zhou et al. introduced a novel method for the production of continuously twisted nanofiber yarn using stepped airflow electrospinning and friction twisting together (Table 3.1I). PAN, PVDF, and PU nanofiber yarns were produced with rates of 4.207, 3.689, and 3.891 g/h. SEM investigations and strength tests have shown that nanofibers were aligned in nanoyarn structure perfectly; nanoyarns performed better than ordinary commercial viscose filament in terms of the stress and strain at break making them a good candidate for weaving applications [43]. Kangazi et al. produced glass nanofibrous yarn containing silver nanoparticles via electrospinning in 2018 (Table 3.1J). The nanofibrous strand was twisted to form the yarn. The tensile strength of nanofibrous yarn was found as 2000 turns per meter (TPM). The diameter of the nanofibers in the yarn structure was measured as 136 nm approximately [44]. In 2018, Demir et al. produced PAN-based carbon nanofiber yarn using a rotating funnel and wind-up unit along with electrospinning (Table 3.1K). Nanofibrous yarns were drawn from the collector to the winding unit with the guidance of a conventional yarn. The average diameter of nanofiber yarn was found between 210 nm and 340 nm and decreased to between 160 nm and 300 nm after carbonization [45]. Jin et al. developed a novel dynamic rotating electrospinning method and used the spinning device for the fabrication of nanofibrous yarn from polysulfone amide (PSA) (Table 3.1L) in 2019. The twist of nanoyarns increased upon increasing the rotation speed of the funnel-shaped collector while mechanical properties decreased. Nanoyarn diameter increased with increasing solution concentration [46]. Yang et al. developed a hybrid strategy for the production of composite yarns made of nanofibers and microfibers together in 2019 (Table 3.1M). The fluorescent tracer technique was used for the structural characterization of nanofiber yarns. Authors reported that composite yarns from nanofiber/microfiber provide durability for preparation, daily use, and washing. It was also reported that developed nanoyarns are ideal for knitting, weaving, and braiding applications [47]. In 2020, Mülayim and Göktepe produced electrospun PAN nanofiber yarn using a conical collector (Table 3.1N). The ideal collector speed was found between 400 and 500 rpm for feasible yarn properties and spinning efficiency. Well-structured twisted yarns were obtained and the system potentially provided an ideal bicomponent electrospun nanofiber yarn formation. In the end, the first fabric sample with nanofiber yarns was woven manually and its characteristics were analyzed [48]. Wang et al. introduced a method for manufacturing conductive nanofiber yarn using dry electrospinning offering self-triggered thread formation in 2020 (Table 3.1O). Instead of mechanical components, the twisting process is triggered by the current electrospinning system as reported by the authors. The longer and more uniformly twisted nanofiber yarns were obtained due to the extended processing time. The system also reduced the electricity consumption that is required for nanoyarn production enabling a cost-efficient method [49]. Levitt et al. developed continuous and scalable multifunctional MXene-infiltrated stretchable nanoyarns using bath electrospinning (Table 3.1P). The stretchability of nanoyarns was achieved up to 43% by varying the flake size and MXene concentration. SEM investigations revealed that some nanoyarns had a ribbon-like cross-section while others had a circular cross-section depending on solution properties. Authors reported that these nanoyarns possess a potential for stretchable electronics and body movement monitoring applications [50].

Apart from pure nanoyarn studies, a few studies introduced novel hybrid methods for the preparation of nanofiber-covered composite yarns in which nanofibers were placed as an envelope while traditional yarns were placed as a core element. These composite yarns sound promising for many applications [51–53].

To summarize, during the past 15 years, studies related to nanoyarn production were mostly focused on the development of new setups and durable nanoyarns that will endure the weaving, knitting, and braiding processes.

3.3.2 Nanoyarns via Centrifugal Spinning

Nanofiber yarns via centrifugal spinning were introduced by FibeRio Corporation along with their patent issued in 2009. In fact, this system was used for the first time to produce microscale glass fibers in the 1950s. Since 2009, only a limited number of studies were carried out regarding nanoyarn production using this technique. The studies related to nanoyarn production via centrifugal spinning were presented in Table 3.2 briefly.

Table 3.2 Nanoyarns via centrifugal spinning

Code	Setup	Yarn Images	Principle/ Yarn Properties in Brief	Ref.
A	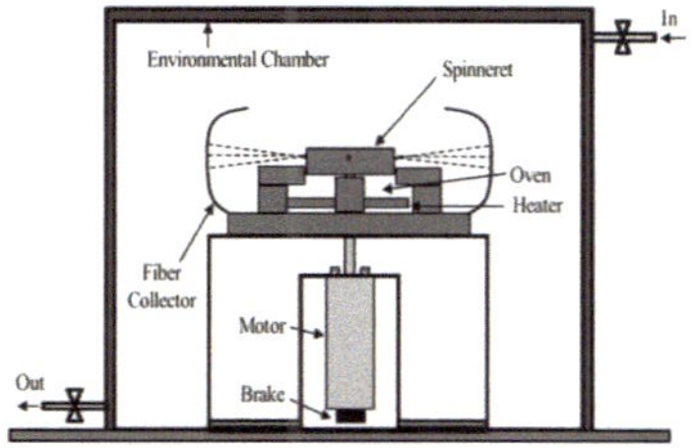	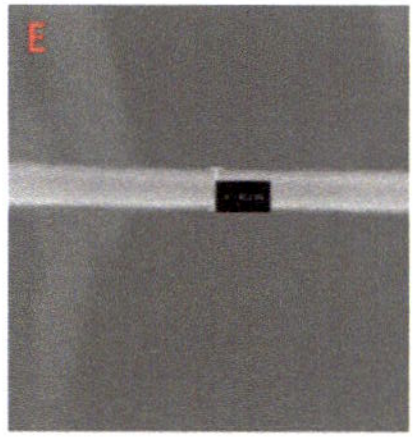	Centrifugal forces. 3000 to 9000 rpm rotational speeds. 220 nm fiber diameters approx.	[54]
B	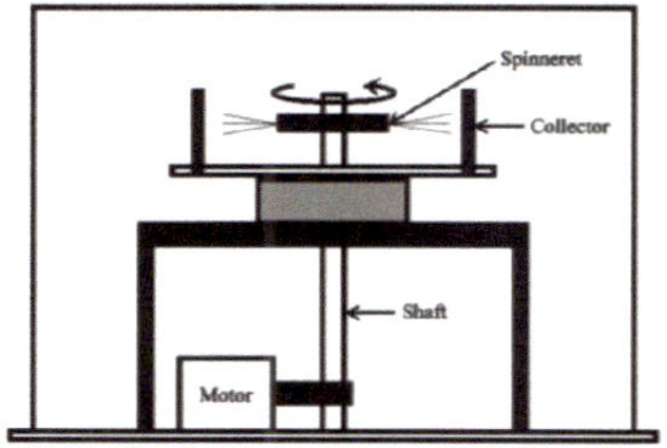	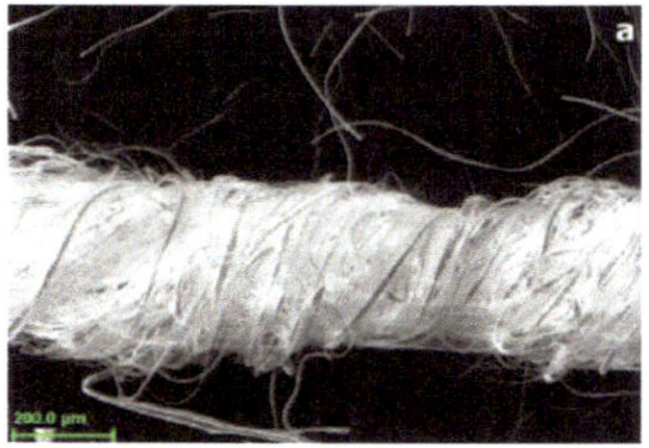	High throughput. Uniform fiber size distribution. Nanoyarn as a protective agent. Commercial potential.	[55]
C	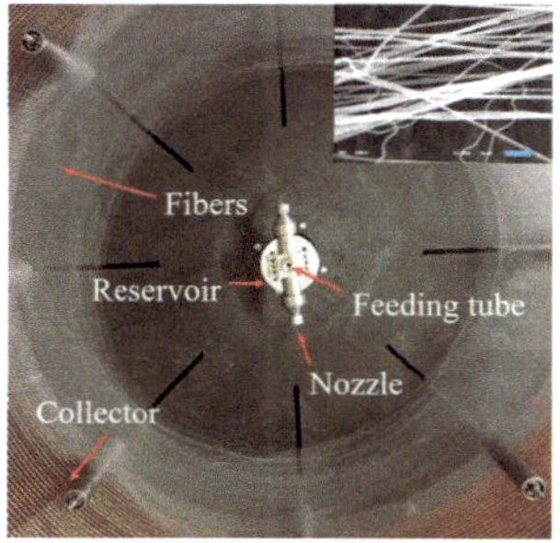	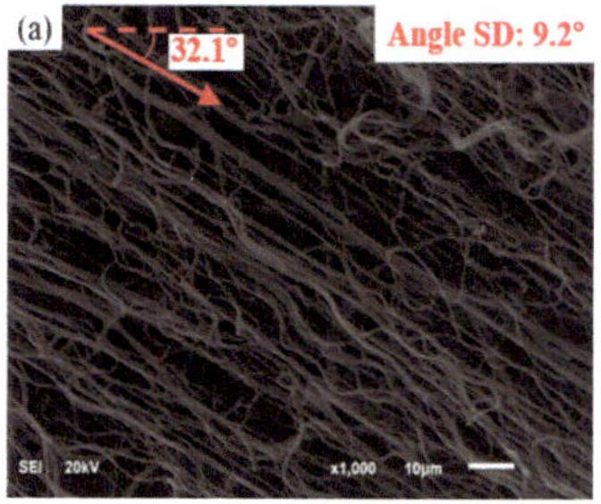	Centrifugal spinning: 3000–5000 rpm. Solvent efficiency for denser polymers. 0.5 mm yarn diameter. Strong enough for cell-culture	[56]

In 2011, McEachin and Lozano produced polycaprolactone nanofibers via Forcespinning^{TM} technology using centrifugal spinning principles (Table 3.2A). Centrifugal spinning uses centrifugal forces to elongate polymer jets so that they can be transformed into nanofibers. The rotational speeds in the centrifugal spinning process varied from 3000 to 9000 rpm as reported by the authors. SEM investigations were carried out and the average diameter of the nanofibers was measured as 220 nm approximately with nearly bead-less mat morphology. The authors stated that uniformly deposited fiber mats can be aligned and spooled into yarns using this technique [54]. Rane et al. produced superhydrophobic Teflon submicron fibers and yarns using the Forcespinning^{TM} technique based on centrifugal forces (Table 3.2B) in 2013. Fibrous samples were processed into nanoyarn form with varying diameters and degrees of twist. SEM investigations revealed that nanofibers had uniform fiber size distribution with ideal length. The authors stated that developed nanofiber yarns are ideal for widespread applications as a protective agent and have commercial potential due to high throughput [55]. In 2019, Li et al. fabricated chitosan composite nanofibers using centrifugal spinning at rotational speeds between 3000–5000 rpm, and then nanofibers were twisted into a yarn form (Table 3.2C). Yarns had 0.5 mm diameter with uniform alignment. Authors reported that denser polymers can be spun with less solvent in comparison to electrospinning making this technique healthier, especially in the case of working with harmful solvents. It was also reported that nanofiber yarns produced from the current technique performed sufficient tensile strength cell culture [56].

To date, the setups that have been used for centrifugal nanofiber spinning didn't change remarkably. Therefore, more attention is paid to existing process parameters and final product properties. Even though centrifugal spinning offers advantages for nanoyarn production in terms of high throughput and possibilities of working with a wide range of polymers, there is still a limited number of studies available in this area.

3.3.3 Nanoyarns via Solution Blowing (Air-Blowing)

Nanoyarn production via solution blowing (air-blowing) is a relatively new technique in comparison to electrospinning and centrifugal spinning. The studies related to nanoyarn production via solution blow spinning were presented in Table 3.3 briefly.

In 2009, Medeiros et al. developed a solution blow spinning (SBS) technique using elements of both electrospinning and melt blowing technologies from poly(methyl methacrylate) (PMMA), polystyrene (PS) and poly(lactic acid) (PLA) polymers (Table 3.3A). In this method, the polymer solution is pumped through an inner nozzle using a syringe pump and delivered through the outer nozzle via high-velocity gas flow, finally producing nanofibers deposited on the collector plate. The authors reported that polymer type and concentration had a greater effect on fiber diameter than other parameters did. Nanofibers were transformed into yarn form using a positioning barrier in front of the nozzle apparatus. It was also stated by the authors that this technique is quite a suitable one for commercial scale-up of nanofiber and yarn production [57]. Jia et al. developed an aligned carbon nanofiber yarn using a modified solution-blowing process in 2013 (Table 3.3B). A pair of parallel rods were used as a collector. Nanofibers of 187 nm diameter were well-aligned in the yarn structure. Yarn performed high conductivity and superior electrochemical properties. Obtained carbon nanofiber yarns can be used as high-performance supercapacitors due to these features [58]. In 2014, Zhuang et al. used a facile spinning-based process for solution blowing of continuous carbon nanofiber yarn (Table 3.3C). A Nanofiber collection system was specifically designed for this study which included a rotary disk, a thread guide tube, and a bobbin. Funnel shape nanofiber nets were formed and continuously wrapped into the yarn that was wound onto the bobbin. According to the SEM observations, nanofibers in yarn structure had 136 nm fiber diameter after the carbonization process. With its high conductivity and electrochemical performance, carbon nanofiber yarn is promising as a one-dimensional supercapacitor electrode [59]. Yang et al. developed solution-blown aligned nanofiber yarn using a modified solution-blowing process (Table 3.3D) in 2020. To obtain continuously aligned nanofiber yarn, a funnel and a turntable as a collector were used. SEM observations revealed that nanofiber yarns were well-aligned and twisted. The authors stated that this method could revolutionize nanofiber production into a mass form. Nanofiber yarns exhibited high performance as a supercapacitor and authors reported that a light-emitting diode can be enlighted by an 8-cm long yarn-shaped supercapacitor for 2 minutes approximately [60].

Table 3.3 Nanoyarns via solution blow spinning

Code	Setup	Yarn Images	Principle/Yarn Properties in Brief	Ref.
A	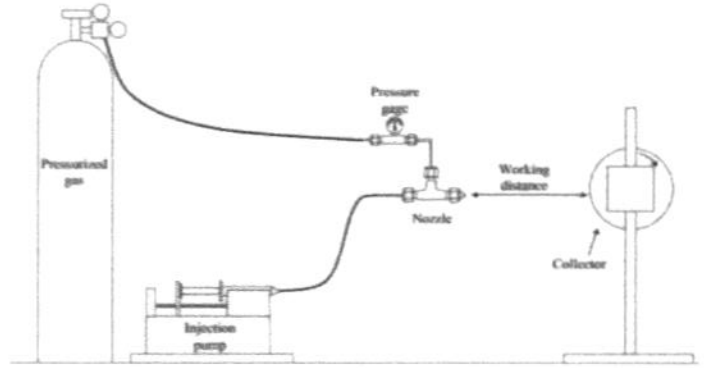	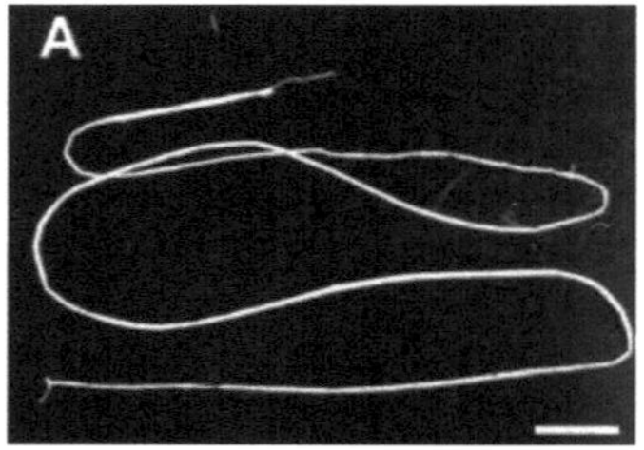	Hybrid solution blowing: electro-spinning + meltblown. Positioning barrier for yarn formation. Commercial scale-up potential.	[57]
B	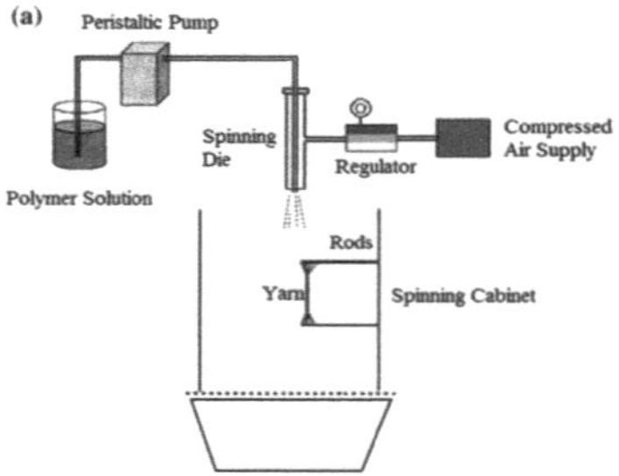	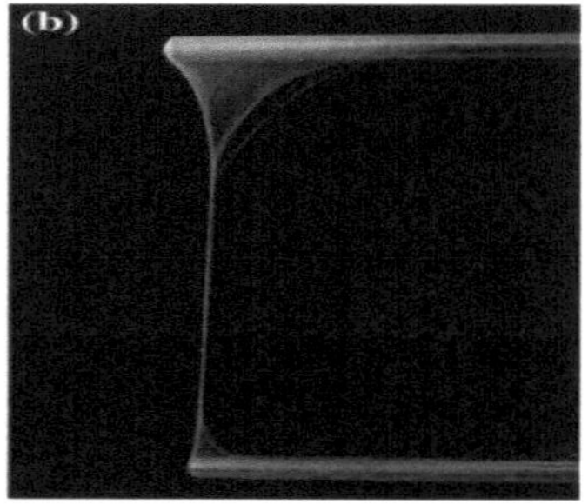	Modified SBS. Parallel rod collector. Highly conductive yarn. Supercapacitor potential.	[58]

Code	Setup	Yarn Images	Principle/Yarn Properties in Brief	Ref.
C	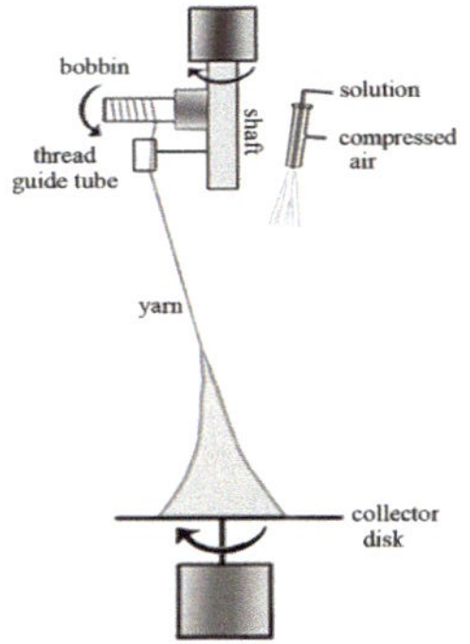		Modified setup & facile based spinning. Funnel shape nanofiber net. Aligned nanofiber yarn. One-dimensional supercapacitor electrode.	[59]
D	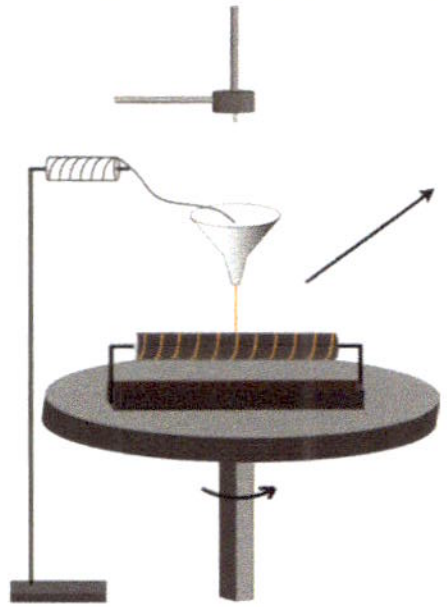	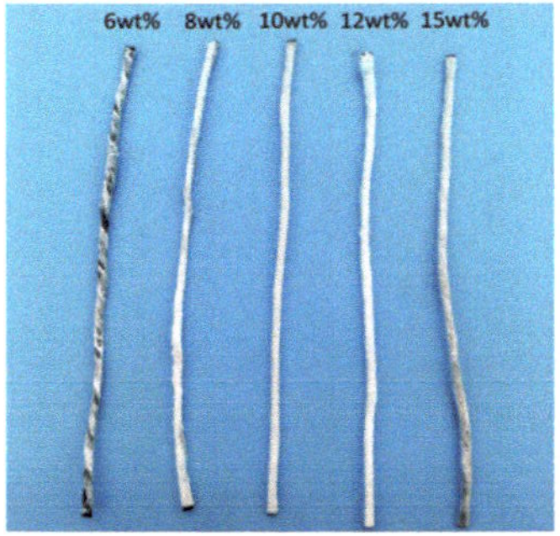	A funnel and a turntable as collector. Aligned carbon nanofiber yarn. Mass production of nanofiber yarn. Supercapacitor.	[60]

Past studies have shown that different setups for nanofiber yarn production via solution blow spinning were introduced by a few research groups with modifications specific to each study. The SBS technique is promising in terms of the mass production of nanofiber yarns commercially.

3.4 Applications of Nanoyarns

Nanoyarns find uses in many areas from tissue engineering to drug delivery, energy, and sensory applications due to superior properties such as large surface area, regeneration, improved delivery, power generation, and sensing. It was discussed in detail for different application areas in subsections.

3.4.1 Nanoyarns in Tissue Engineering

Nanoyarns have been widely used in the tissue engineering area in the last decade. Cell culture, scaffold building, tissue repair, and regeneration are the main activities within this area where nanoyarns are making a considerable contribution.

In 2012, Wu et al. prepared a nanoyarn scaffold from silk fibroin (SF)/poly(L-lactide-co-caprolactone) P(LLA-CL) using dynamic liquid electrospinning. A water vortex was used for nanofiber deposition and a rotating mandrel was used for the collection of nanoyarns. Researchers cultured mouse fibroblasts on the nanoyarn scaffolds. SEM observations showed that the nanoyarn scaffold has excellent surface properties and a porous structure. Nanoyarn scaffolds accelerated proliferation, organized morphology, and improved infiltration of cells in comparison to traditional nanofibrous scaffolds [61]. Wu et al. developed a novel collagen/P(LLA-CL) nanoyarn scaffold using dynamic electrospinning for cell infiltration and vascularization in 2013. Nanoyarns with capillary-like formations formed *in vitro* guided cell behaviors. The biomimetic structure of nanofiber yarn promoted cell adhesion and proliferation. These biomimetic nanoyarns were reported as potential scaffolds for use in tissue engineering [62]. In 2013, Xu et al. fabricated a poly(L-lactide-co-ε-caprolactone)/collagen nanoyarn network using electrospinning for tendon tissue engineering. Nanoyarns had superior morphological properties such as a three-dimensional, macroporous, and aligned structure that mimics the extracellular matrix of native tendon tissues. The nanoyarn scaffold had a 641 ± 68 nm fiber diameter. The spatial structure of the nanoyarn scaffold promoted cell infiltration and mechanical properties were found promising for tendon tissue regeneration [63]. Liu et al. developed nanofiber yarn bone regeneration using electrospinning with a dynamic liquid support system in 2013. Nanoyarns were chopped into short nanoyarns and just after they were incorporated in Type I collagen hydrogel. Collagen incorporation improved the mechanical properties of nanoyarns. The proliferation of human mesenchymal stem cells was realized efficiently according to the *in vitro* study. The authors stated that nanofiber yarns possess great potential in bone regeneration [64]. In 2014, Yang et al. fabricated a nanoyarn-reinforced nanofibrous scaffold for tendon tissue engineering via electrospinning using a two-collector system. Nanoyarn reinforcement improved mechanical properties that are beneficial for tendon tissue regeneration. This improvement advanced cell proliferation and infiltration enabling an ideal nanoyarn for tendon tissue engineering applications [65]. Sun et al. fabricated mineralized P(LLA-CL)/SF three-dimensional nanoyarn scaffolds using a dynamic liquid support electrospinning system for bone tissue regeneration in 2015. SEM investigations showed that yarn in the nanoyarn scaffold was twisted by many nanofibers. Infiltration was improved due to the porous structure of the nanoyarn scaffold. Biomineralization improved the proliferation of cells and the osteoconductivity of nanoyarn scaffolds which increased the efficiency of bone tissue engineering [66]. Following these works, several studies were carried out in the tissue engineering area using the electrospinning technique between 2016–2021 years. These works covered different functional use of nanoyarns such as nanoyarns for the application of urethroplasty, nanoyarn nerve guidance conduits for nerve tissue engineering, nanoyarn for annulus fibrosus tissue engineering, nanofiber yarn enabling highly stretchable engineered microtissue, nanofiber yarn for anisotropic tissue regeneration and nanoyarn for stem cell-based tissue regeneration. In these studies, nanoyarns provided various benefits for tissue engineering such as improvement of biocompatibility, mechanical properties, and biological degradability, enhanced tissue repair and regeneration, successful tissue replacement, functionalized highly stretchable microtissues, anisotropic tissue regeneration and biomechanically improved scaffolds [67–72]. In addition to nanoyarn for tissue engineering developed via electrospinning, centrifugal spinning was also used to develop nanoyarn for the tissue engineering area and these studies were already introduced in previous sections.

3.4.2 Nanoyarns in Drug Delivery

Nanofibers offer an advanced platform for the delivery of drugs and other biological agents. The type of nanofibrous scaffolds dramatically affects the delivery action [73]. Therefore, the process parameters and morphology should be designed and studied considering these facts. In 2016, Padmakumar et al. developed polymeric core-sheath yarns as drug-eluting surgical sutures using electrospinning. The authors stated that as a functional material, the drug-laden coating provides good, localized drug concentration while variable loading efficiency and release kinetics limit its use. Therefore, novel materials are needed to advance the drug delivery process eliminating limitations. Mechanical tests and release analysis showed that electrospun yarn performed as mechanically strong, providing prolonged drug release [74]. Bae et al. developed heparin-eluting electrospun nanofiber yarns for antithrombotic vascular sutures. SEM investigations revealed that electrospun nanofiber yarns had highly uniform morphology with a smooth surface. The average diameter of nanoyarn was measured as 100 μm approximately making this yarn an ideal microvascular suture. Authors reported that the diameter of the yarns can be easily adjusted by increasing or decreasing the spinning time and delivery action can be tailored in this way [75]. Padmakumar et al. developed woven polymeric nanotextiles from implantable electrospun polydioxanone (PDS) nanoyarns providing long-term drug delivery in 2019. Nanoyarns were fabricated by a modified electrospinning setup and a hemispherical rotating collector was used. The nanoyarn-based advanced structure of textile enabled tunable drug delivery [76]. In this section, studies into nanoyarns for drug delivery applications via electrospinning were introduced as no available information has been found for centrifugal spun and/ or solution-blown nanoyarns in the relevant literature.

3.4.3 Nanoyarns in Energy

Nanofibers have been used commonly in energy production, storage, and conversion applications in the last decade [77]. In 2016, Rahbar et al. fabricated nanofiber yarns including microencapsulated phase change materials from nylon 6 for thermal regulation activity using electrospinning. SEM results revealed that nanoyarns had good morphology with a smooth surface. Differential scanning calorimetry analysis produced successful results and authors stated that these phase change nanofiber yarns are promising for thermal energy storage and regulation activities [78]. In 2019, Gao et al. fabricated core-sheath nanoyarn for wearable piezoelectric nanogenerator application via the touch spinning method where nanofibers are produced by the direct drawing of polymer solutions or melt using mechanical force. Authors reported that 0.72 V electrical potential difference can be harvested under compression of 0.33 MPa for a single 3-cm long yarn. It was also reported that this novel nanoyarn-based piezoelectric nanogenerator (PNG) is capable to convert the mechanical energy of body motion into electricity and promising for energy-harvesting smart textiles [79]. In 2020, Ma et al. manufactured hybridized nano–micro triboelectric yarns for energy harvesting and signal sensing using electrospinning. SEM investigations revealed that PVDF and PAN hybrid nanofibers uniformly wrapped around the conductive yarn. It was stated that the diameter of the yarn can be controlled by adjusting the forward speed of the winder. The average diameter of nano–micro hybridized core–shell structured yarn was measured as 350.66 μm. Authors reported that this yarn can be used in the mass production of smart textiles with an energy-harvesting function [80]. In brief, nanoyarns extended the use of nanofibers in durable applications while enhancing functional capabilities. In this section, only studies into nanoyarns for energy applications via electrospinning were introduced as was already mentioned about solution-blown nanoyarns for energy area in previous sections. No available information has been found for centrifugally spun nanoyarns for the energy applications area.

3.4.4 Nanoyarns as Sensors

Nanoyarns inherit superior properties of nanofibers in terms of high precision, improved sensing, and tailored conductivity and enhance their functionalities thanks to their controllable nature. Therefore, many researchers developed nanoyarn forms of these materials to benefit the aforementioned properties to maximize efficiency in functional applications.

In 2016, Liu et al. developed a fast-response ammonia sensor based on coaxial PPy–PAN nanofiber yarn using a novel modified electrospinning setup. SEM analysis showed that nanofibers had a highly oriented alignment in the one-dimensional structure of nanofiber yarn.

The average diameter of the nanofiber yarn was measured as 105 µm. The advanced sensing capabilities of the nanofiber yarn sensor make it an ideal candidate for the intelligent electronics area [81]. Wu et al. fabricated flexible and conductive nanofiber-structured single yarn sensors for smart wearable devices using a novel electrospinning method in 2017. SEM images showed that a highly oriented nanofiber arrangement in nanoyarn structure was achieved. This arrangement provided excellent sensitivity and fast response/recovery capabilities to the nanoyarn. Due to the coaxial nature of the nanoyarn, high mechanical performance and high conductivity were obtained at the same time. Authors reported that this nanoyarn can be used as a sensing element for wearable gas sensors [82]. Zhou et al. developed highly sensitive, self-powered, and wearable electronic skin based on a pressure-sensitive fabric sensor using electrospinning in 2017. Nanofibers in yarn structure had a diameter ranging from 100 nm to 200 nm. Nanoyarns were then woven into fabric form having a multi-leveled hierarchical structure that provided a high-pressure sensitivity. Authors reported that electrospun nanoyarn has the potential for use as a wearable pressure sensor [83]. Following these works, a few more studies using electrospinning were carried out by different research groups into nanoyarns for sensor areas between 2018–2020 years. These works covered novel studies such as the development of highly sensitive and stretchable fabric sensors for human health monitoring, high-pressure sensitivity and spatial tactile acuity for smart textiles, all-organic, conductive, and biodegradable yarn sensors for fabrication of a capacitor device, and nanoyarn-based multifunctional sensors for soft electronics [84–87]. No available information has been found for centrifugal spun or solution-blown nanoyarns for energy applications area.

3.4.5 Future Potential Applications

Apart from applications mentioned in previous sections, nanoyarns have been used in nanofluidic applications for instance they can be used as probes working just like a sponge drawing up fluids from microcapillaries, small organisms, and, ideally, from a single cell [88]. Nanoyarns may find more common use in this kind of area in the future. Threads in surgical operations are an ever-relevant topic and nanoyarns are promising especially ones combining biodegradability and performance. Nanoyarns may also find uses in the dentistry area, especially by enhancing the surface area and alleviating stresses at the bone-implant interface during dental implant operations [89]. Construction applications can be a potential area and nanoyarns can be used as a reinforcement element in matrixes of composites. The personal protective equipment (PPE) area became an area of interest, especially during the Covid-19 pandemic era and it is expected that PPE products and R&D in this area will continue to be relevant in the post-pandemic era. Nanoyarns have a great potential for improvement of performance and protective properties of PPE items with their aligned, functionalized, and durable structure. The filtration area is becoming more competitive in recent years, woven fabric filters from nanoyarns could boost usage and filtration performance especially pressure drops can be minimized thanks to the superior properties of nanoyarns. Potential applications are only limited by imagination while considering feasibility, cost, and consumer preferences.

3.5 Future Prospects in Nanoyarns

As was stressed in previous sections, nanoyarns are mostly produced by electrospinning followed by solution blow spinning and centrifugal spinning, respectively. In the future, modifications of these techniques may extend the development possibilities of nanoyarns. Especially hybrid methods are of interest in recent years for nanoyarns production and probably new modifications will be carried out in this area to develop the next generation of nanoyarn in the future. Electrically assisted solution blowing (electroblowing) is one of these hybrid methods. In this method, the polymer jet is exposed to electrical and air shear forces, and fibers formed in this way (Fig. 3.1). As this hybrid method combines the advantages of two methods, nanoyarns with regular nanofibers and a high production rate could be produced, therefore this method has a great potential in scale-up of production of nanoyarns in future [90].

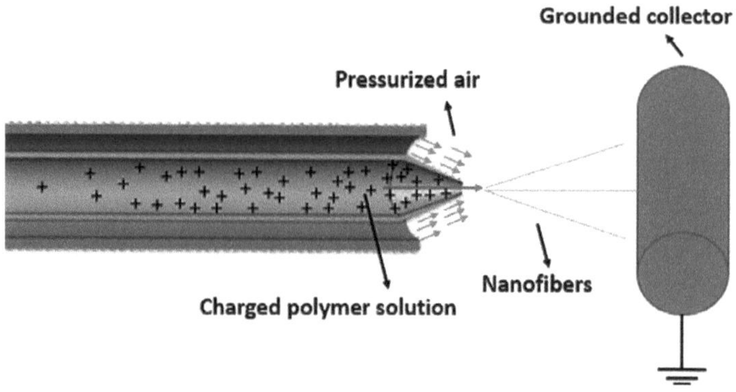

Figure 3.1 Schematic illustration of electroblowing [90].

Centrifugal electrospinning is another hybrid method combining the most common methods for nanofiber production in which electrospinning is the method producing ultrafine nanofibers with high performance and centrifugal spinning with high fiber production rates (Fig. 3.2) [91].

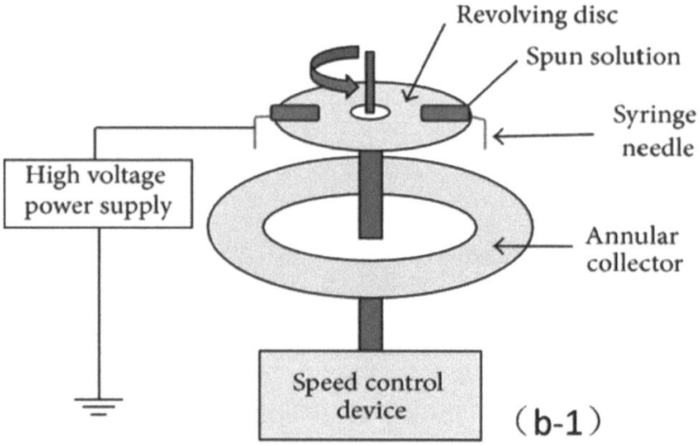

Figure 3.2 Schematic of the horizontal centrifugal electrospinning equipment [91, 92].

Working with a wide range of polymers is another advantage of centrifugal electrospinning thanks to centrifugal forces and nanoyarns with various functionalities from different raw materials can be fabricated in this way [93].

On the other hand, the Bubbfil spinning process combines electrostatic force, blowing air, or mechanical force to overcome the surface tension and it mainly includes the bubble electrospinning or blown bubble spinning (Fig. 3.3) [94].

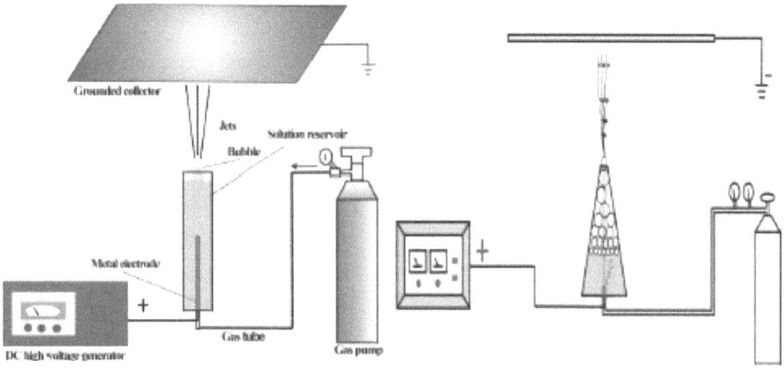

Figure 3.3 Bubble electrospinning setup for tunable bubble size [94, 95].

High-temperature and high-speed airflow to directly blow and stretch polymer multi-bubbles are utilized in blown bubble spinning to produce nanofibers. This technique could be used to fabricate hierarchically structured nanofibrous yarns and may become an important alternative in the mass production of nanoyarns in the future [96].

Electrocarding is another promising technology for production of mass nanoyarns and the mass production of nanofabrics from nanoyarns which combines electrospinning and conventional yarn spinning. In electrocarding, electrospun nanofibers are collected in fiber web bundles and they are converted into nano-slivers by fiber drawing, before yarn spinning (Fig. 3.4). After the scale-up process, Electrocarding may become one of the most competitive methods for nanoyarn production in future [97].

Several studies into the development of novel nanoyarn materials have been carried out until now and it is expected to experience an exponential increase in research and commercialization of this area in terms of new setups, new features, and applications [98–100].

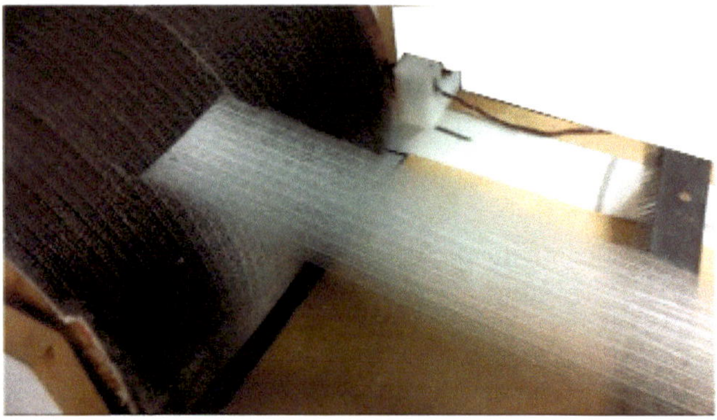

Figure 3.4 The electrocarding of nylon 6 nanofibers into an orderly nano-sliver for yarn spinning [97].

Bicomponent nanofiber yarn technology is promising for future developments of multifunctional textiles, but there were limited numbers of studies available until now whose details were already discussed in previous sections. This technology can be extended to tricomponent and more components to advance development possibilities.

New designs of collectors may open up new possibilities for differentiated alignment and morphology developments of nanoyarns. Especially 3-dimensional printers are low-cost and flexible to carry out this task. Novel 3D shape collectors could be designed using 3D printers and even some advanced designs can provide twisting and/or false twisting processes simultaneously during nanofiber spinning.

Test equipment that is specific to nanoyarns testing can be designed and manufactured. Existing instruments can be used for typical tests such as yarn hairiness or twist count analysis, but it is better to develop a new instrument that is tailored for the tensile properties analysis of nanoyarns. Alternative to this, a new type of clamps can be designed and manufactured for existing tensile testers having a specific configuration for tensile analysis of nanoyarns.

There is no test standard available for nanoyarns. This is an untouched area and offers new development possibilities. New test methods and new standards related to these newly developed methods can be introduced especially for tensile properties, yarn hairiness, and single nanofiber analysis in the nanoyarn structure.

Academy and industry collaboration is a must to achieve this goal as the academy is pioneering in this area for inventions however industry plays an important role in field trials, scale-up, funding projects, and commercialization.

3.6 Conclusion

The nanoyarn area is quite prone to new developments in terms of material properties for end uses, new applications, new instruments, and set up designs with patenting and/or issuing a utility model potential. In conclusion, material science, the invention of novel raw materials, and innovative instruments will drive the development of new products and applications in this area.

References

1. Najafi, M. and Frey, M.W. (2021). Electrospun nanofibers for chemical separation. *Nanomaterials*, 2020, 10, 982; doi:10.3390/nano10050982.
2. Hussain, M.M. and Ramkumar, S.S. (2006). Functionalized nanofibers for advanced applications. *Indian Journal of Fiber and Textile Research*, 31, 41–51.
3. Formhals, A. (1934). Process and apparatus for preparing artificial threads. In: Google Patents.
4. Koenig, K., Beukenberg, K., Langensiepen, F. and Seide, G. (2019). A new prototype melt-electrospinning device for the production of biobased thermoplastic sub-microfibers and nanofibers. *Biomaterials Research*, 23, article no. 10.
5. Barhoum, A., Rasouli, R., Yousefzadeh, M., Rahier, H. and Bechelany, M. (2018). Nanofiber technology: History and developments. In: Barhoum, M. and Bechelanyand A. Makhlouf (eds.), *Handbook of Nanofibers* (pp. 1–42). Switzerland: Springer International Publishing AG. https://doi.org/10.1007/978-3-319-42789-8_54-1.
6. Gilbert, W. and Wright, E. (1967). *De magnete, magneticisque corporibus, et de magno magnete tellure: physiologia noua, plurimis and argumentis and experimentis demonstrata.* Londini: excudebat Short.
7. Bose G-M. (1744). *Die Electricität nach ihrer Entdeckung und Fortgang mit poetischer Feder entworffen.* Wittenberg: Joh. Joachim Ahlfelden.

8. Rayleigh, L. (1878). The influence of electricity on colliding water drops. *Proceedings of the Royal Society of London*, 28, 405–409.
9. Luo, C.J., Stoyanov, S.D., Stride, E., Pelanb, E. and Edirisinghe, M. (2012). Electrospinning versus fibre production methods: From specifics to technological convergence, *Chemical Society Reviews*, 41, 4708.
10. Lin, T., Wang, H. and Wang, X. (2005). Self-crimping bicomponent nanofibers electrospun from polyacrilonitrile and elastomeric polyurethane. *Advanced Materials*, 17(22), 2699–2703.
11. Chen, S., Hou, H., Hu, P., Wendorff, J., Greiner, A. and Agarwal, S. (2009). Polymeric nanosprings by bicomponent electrospinning. *Macromolecular Materials and Engineering*, 294, 265–271.
12. Yang, D., Li, Y. and Nie, J. (2007). Preparation of gelatin/PVA nanofibers and their potential application in controlled release of drugs. *Carbohydrate Polymers*, 69(3), 538–543.
13. Song, C. and Dong, X. (2012). Preparation and characterization of tricomponent $SiO_2/SnO_2/TiO_2$ composite nanofibers by electrospinning. *Optoelectronic and Advanced Materials*, 6(1–2), 225–229.
14. Kim, T., G., Chung, H., J. and Park, T. (2008). Macroporous and nanofibrous hyaluronic acid/collagen hybrid scaffold fabricated by concurrent electrospinning and deposition/leaching of salt particles. *Acta Biomaterialia*, 4(6), 1611–1619.
15. Bonino, C.A., Efimenko, K., Jeong, S.I., Krebs, M.D., Alsberg, E. and Khan, S.A. (2012). Three-dimensional electrospun alginate nanofiber mats via tailored charge repulsions. *Small*, 8(12), 1928–1936.
16. Sivri, Ç. (2017). Development of nanofiber structures having novel morphologies differentiated with the help of organic acids. *Journal of the Textile Institute*, 108, 1879–1887.
17. Yousefzadeh, M., Latifi, M., Tehran, M.A., Teo, W.E. and Ramakrishna, S. (2012). A note on the 3D structural design of electrospun nanofibers. *Journal of Engineered Fabrics and Fibers*, 7(2), 17–23.
18. Blakeney, B.A., Tambralli, A., Anderson, J.M., Andukuri, Lim, D.J., Dean, D.R. and Jun, H.W. (2011). Cell infiltration and growth in a low density, uncompressed three-dimensional electrospun nanofibrous scaffolds. *Biomaterials*, 32(6), 1583–1590.
19. He, J.H., Liu, Y., Xu, L., Yu, Y.J. and Sun, G. (2008). Biomimic fabrication of electrospun nanofibers with high throughput. *Chaos, Solitons and Fractals*, 37(3), 643–651.

20. Yan, G., Yu, J., Qiu, Y., Yi, X., Lu, J., Zhou, X. and Bai, X. (2011). Self-assembly of electrospun polymer nanofibers: a general phenomenon generating honeycomb-patterned nanofibrous structures. *Langmuir*, 27(8), 4285–4289.

21. Sarkar, K., Gómez, C., Zambrano, S., Ramirez, M., Hoyos, E., Vasquez, H. and Lozano, K. (2010). Electrospinning to Forcespinning^{TM}. *Materials Today*, 13, 12–14. 10.1016/S1369-7021(10)70199-1.

22. Zhang, X. and Lu, Y. (2014). Centrifugal spinning: an alternative approach to fabricate nanofibers at high speed and low cost. *Polymer Reviews*, 54, 677–701.

23. Lozano, K. and Sarkar, K. Inventors: Superfine fiber creating spinneret and uses thereof, US Patent 2009/0232920 A1, 2009.

24. Edmondson, D., Cooper, A., Jana, S., Wood, D.M. and Zhang, M. (2012). Centrifugal electrospinning of highly aligned polymer nanofibers over a large area. *Journal of Materials Chemistry*, 22, 18646–18652.

25. Shuliang, L., Long, Y., Zhang, Z., Zhang, H., Sun, B., Zhang, J. and Han, W. (2013). Assembly of oriented ultrafine polymer fibers by centrifugal electrospinning. *Journal of Nanomaterials*, 2013, 8. https://doi.org/10.1155/2013/713275

26. Dadol, G.C., Kilic, A., Tijing, L., Lim, K.J., Cabatingan, L.K., Tan, N., Stojanovska, E. and Polat, Y. (2020). Solution blow spinning (SBS) and SBS-spun nanofibers: Materials, methods, and applications. *Materials Today Communications*, 25, 101656.

27. Ali, U., Zhou, Y., Wang X. and Lin, T. (2011). Electrospinning of Continuous Nanofiber Bundles and Twisted Nanofiber Yarns, In: Dr. Tong Lin (ed.), *Nanofibers: Production, Properties and Functional Applications*, ISBN: 978-953-307-420-7, InTech.

28. Liu, C.K., Sun, R.J., Lai, K., Sun, C.Q. and Wang, Y.W. (2008). Preparation of short submicron-fiber yarn by an annular collector through electrospinning. *Materials Letters*, 62, 4467–4469.

29. Dalton, P.D., Klee, D. and Mãller, M. (2005). Electrospinning with dual collection rings. *Polymer*, 46, 611–614.

30. Sanatgar, R.H., Borhani, S., Ravandi, S.A. and Gharehaghaji, A.A. (2012). The influence of solvent type and polymer concentration on the physical properties of solid state polymerized PA66 nanofiber yarn. *Journal of Applied Polymer Science*, 126, 1112–1120.

31. Abbasipour, M. and Khajavi, R. (2013). Nanofiber bundles and yarns production by electrospinning: A review. *Advances in Polymer Technology*, 32, 21363.

32. Jin, S., Xin, B., Zheng, Y. and Liu, S. (2018). Effect of electric field on the directly electrospun nanofiber yarns: Simulation and experimental study. *Fibers and Polymers*, 19, 116–124.
33. Moghbelnejad, Z., Gharehaghaji A.A., Yousefzadeh, M. and Hajiani, F. (2021). Investigation of wicking phenomenon and tensile properties in three-layer composite nanofibrous PA/PLLA yarn. *Polymer Engineering and Science*, 61, 576–585. https://doi.org/10.1002/pen.2560.
34. Göktepe, F. and Mülayim, B.B. (2018). Long path towards to success in electrospun nanofiber yarn production since 1930s: A critical review. *Autex Research Journal*, 18, 109–87.
35. Smit, E., Buttner, U. and Sanderson, R.D. (2005). Continuous yarns from electrospun fibers. *Polymer*, 46(8), 2419–2423. https://doi.org/10.1016/j.polymer.2005.02.002.
36. Wang, X., Zhang, K., Zhu, M., Yu, H., Zhou, Z., Chen, Y. and Hsiao, B.S. (2008). Continuous polymer nanofiber yarns prepared by self-bundling electrospinning method. *Polymer*, 49, 2755–2761.
37. Bazbouz, M.B. and Stylios, G.K. (2012). A new mechanism for the electrospinning of nanoyarns. *Journal of Applied Polymer Science*, 124, 195–201.
38. Yousefzadeh, M., Latifi, M., Teo, W.-E., Amani-Tehran, M. and Ramakrishna, S. (2011). Producing continuous twisted yarn from well-aligned nanofibers by water vortex. *Polymer Engineering and Science*, 51(2), 323–329. https://doi.org/10.1002/pen.21800.
39. Chvojka, J., Hinestroza, J.P. and Lukas, D. (2013). Production of poly(vinylalcohol) nanoyarns using a special saw-like collector. *Fibres and Textiles in Eastern Europe 21*, 2(98), 28–31.
40. Shuakat, M.N. and Lin, T. (2015). Highly twisted, continuous nanofiber yarns prepared by a hybrid needle-needleless electrospinning technique. *RSC Advances*, 5, 33930–33937.
41. Levitt, A.S., Knittel, C.E., Vallett, R., Koerner, M., Dion, G. and Schauer, C.L. (2017). Investigation of nanoyarn preparation by modified electrospinning setup. *Journal of Applied Polymer Science*, 134(19), 44813.
42. Mehrpouya, F., Foroughi, J., Naficy, S., Razal, J. and Naebe, M. (2017). Nanostructured electrospun hybrid graphene/polyacrylonitrile yarns. *Nanomaterials*, 7(10), 293.
43. Zhou, Y., Wang, H., He, J., Qi, K., Ding, B. and Cui, S. (2018). Novel method for preparation of continuously twisted nanofiber yarn based on a

combination of stepped airflow electrospinning and friction twisting. *Journal of Materials Science*, 53, 15735–15745.

44. Kangazi, M.K., Gharehaghaji, A.A. and Montazer, M. (2018). Glass nanofibrous yarn through electrospinning along with in situ synthesis of silver nanoparticles. *Journal of Sol-Gel Science and Technology*, 88, 528–540.
45. Demir, A., Acikabak, B., Ahan, Z. (2018). *IOP Conference Series: Materials Science and Engineering*, 460 012027.
46. Jin, S., Xin, B. and Zheng, Y. (2019). Preparation and characterization of polysulfone amide nanoyarns by the dynamic rotating electrospinning method. *Textile Research Journal*, 89(1), 52–62. doi:10.1177/0040517517736474.
47. Yuchen, Y., Yajie, Z., Quan, Z., Hong-nan, Z., Qin, X., Wang R. and Yu, J. (2019). An efficient hybrid strategy for composite yarns of micro-/nano-fibers. *Materials and Design*, 184, 108196.
48. Mülayim, B.B. and Göktepe, F. (2020). Analysis of polyacrylonitrile nanofiber yarn formation in electrospinning by using a conical collector and two oppositely charged nozzles. *Journal of the Textile Institute*, 112, 494–504.
49. Wang, W.C., Cheng, Y.T. and Estroff, B. (2020). Electrostatic self-assembly of composite nanofiber yarn. *Polymers*, 13(1), 12. https://doi.org/10.3390/polym13010012.
50. Levitt, A., Seyedin, S., Zhang, J., Wang, X., Razal, J., Dion, G. and Gogotsi, Y. (2020). Bath electrospinning of continuous and scalable multifunctional MXene-infiltrated nanoyarns. *Small*, 16(26), e2002158.
51. Memiş, N.K., Kayabaşı, G. and Yılmaz, D. (2019). Development of a novel hybrid yarn production process for functional textile products. *Journal of Industrial Textiles*, 48, 1462–1488.
52. Valtera, J., Kalous, T., Pokorný, P., Baťka, O., Bílek, M., Chvojka, J., Mikes, P., Košťáková, E., Žabka, P., Ornstova, J., Beran, J., Stanishevsky, A. and Lukas, D. (2019). Fabrication of dual-functional composite yarns with a nanofibrous envelope using high throughput AC needleless and collectorless electrospinning. *Scientific Reports*, 9, article no. 1801.
53. Yalcinkaya, F., Komárek, M., Lubasová, D., Sanetrník, F. and Maryska, J. (2016). Preparation of antibacterial nanofibre/nanoparticle covered composite yarns. *Journal of Nanomaterials*, 2016, 7565972.
54. McEachin, Z. and Lozano, K. (2012). Production and characterization of polycaprolactone nanofibers via Forcespinning™ technology. *Journal of Applied Polymer Science*, 126, 473–479.

55. Rane, Y., Altecor, A., Bell, N. and Lozano, K. (2013). Preparation of superhydrophobic Teflon® AF 1600 sub-micron fibers and yarns using the Forcespinning™ technique. *Journal of Engineered Fibers and Fabrics*, 8. https://doi.org/10.1177/1558925013008004

56. Li, Z., Mei, S., Dong, Y., She, F. and Kong, L. (2019). High efficiency fabrication of chitosan composite nanofibers with uniform morphology via centrifugal spinning. *Polymers*, 11, 1550.

57. Medeiros, E., Glenn, G., Klamczynski, A., Orts, W. and Mattoso, L. (2009). Solution blow spinning: A new method to produce micro- and nanofibers from polymer solutions. *Journal of Applied Polymer Science*, 113, 2322–2330.

58. Jia, K., Zhuang, X., Cheng, B., Shi, S., Shi, Z. and Zhang, B. (2013). Solution blown aligned carbon nanofiber yarn as supercapacitor electrode. *Journal of Materials Science: Materials in Electronics*, 24, 4769–4773.

59. Zhuang, X., Jia, K., Cheng, B., Feng, X., Shi, S. and Zhang, B. (2014). Solution blowing of continuous carbon nanofiber yarn and its electrochemical performance for supercapacitors. *Chemical Engineering Journal*, 237, 308–311.

60. Yang, J., Mao, Z., Zheng, R., Liu, H. and Shi, L. (2020). Solution-blown aligned nanofiber yarn and its application in yarn-shaped supercapacitor. *Materials*, 13, 3778.

61. Wu, J., Liu, S., He, L., Wang, H., He, C., Fan, C. and Mo, X. (2012). Electrospun nanoyarn scaffold and its application in tissue engineering. *Materials Letters*, 89, 146–149.

62. Wu, J., Huang, C., Liu, W., Yin, A., Chen, W., He, C., Wang, H., Liu, S., Fan, C., Bowlin, G. and Mo, X. (2014). Cell infiltration and vascularization in porous nanoyarn scaffolds prepared by dynamic liquid electrospinning. *Journal of Biomedical Nanotechnology*, 10(4), 603–614.

63. Xu, Y., Wu, J., Wang, H., Li, H., Di, N., Song, L., Li, S., Li, D., Xiang, Y., Liu, W., Mo, X. and Zhou, Q. (2013). Fabrication of electrospun poly(L-lactide-co-ε-caprolactone)/collagen nanoyarn network as a novel, three-dimensional, macroporous, aligned scaffold for tendon tissue engineering. *Tissue Engineering Part C: Methods*, 19 12, 925–936.

64. Liu, W., Zhan, J., Su, Y., Wu, T., Ramakrishna, S., Liao, S. and Mo, X. (2014). Injectable hydrogel incorporating with nanoyarn for bone regeneration. *Journal of Biomaterials Science*, Polymer Edition, 25, 168–180.

65. Yang, C., Deng, G., Chen, W., Ye, X. and Mo, X. (2014). A novel electrospun-aligned nanoyarn-reinforced nanofibrous scaffold for tendon tissue engineering. *Colloids and Surfaces B: Biointerfaces*, 122, 270–276.

66. Sun, B., Li, J., Liu, W., Aqeel, B.M., El-Hamshary, H., Al-Deyab, S. and Mo, X. (2015). Fabrication and characterization of mineralized P(LLA-CL)/SF three-dimensional nanoyarn scaffolds. *Iranian Polymer Journal*, 24, 29–40.

67. Guo, X., Zhang, K., El-aassar, M., Wang, N., El-Hamshary, H., El-Newehy, M., Fu, Q. and Mo, X. (2016). The comparison of the Wnt signaling pathway inhibitor delivered electrospun nanoyarn fabricated with two methods for the application of urethroplasty. *Frontiers of Materials Science*, 10, 346–357.

68. Pan, X., Sun, B. and Mo, X. (2018). Electrospun polypyrrole-coated polycaprolactone nanoyarn nerve guidance conduits for nerve tissue engineering. *Frontiers of Materials Science*, 12, 438–446.

69. Ma, J., He, Y., Liu, X., Chen, W., Wang, A., Lin, C., Mo, X. and Ye, X. (2018). A novel electrospun-aligned nanoyarn/three-dimensional porous nanofibrous hybrid scaffold for annulus fibrosus tissue engineering. *International Journal of Nanomedicine*, 13, 1553–1567.

70. Li, Y., Guo, F., Hao, Y., Gupta, S., Hu, J., Wang, Y., Wang, N., Zhao, Y. and Guo, M. (2019). Helical nanofiber yarn enabling highly stretchable engineered microtissue. *Proceedings of the National Academy of Sciences*, 116, 9245–9250.

71. Li, D., Tao, L., Shen, Y., Sun, B., Xie, X., Ke, Q., Mo, X. and Deng, B. (2020). Fabrication of multilayered nanofiber scaffolds with a highly aligned nanofiber yarn for anisotropic tissue regeneration. *ACS Omega*, 5, 24340–24350.

72. Chen, S., Wang, J., Chen, Y., Mo, X. and Fan, C. (2021). Tenogenic adipose-derived stem cell sheets with nanoyarn scaffolds for tendon regeneration. *Materials Science and Engineering C: Materials for Biological Applications*, 119, 111506.

73. Zhang, Z., Hu, J. and Ma, P. (2012). Nanofiber-based delivery of bioactive agents and stem cells to bone sites. *Advanced Drug Delivery Reviews*, 64 12, 1129–1141.

74. Padmakumar, S., Joseph, J., Neppalli, M.H., Mathew, S.E., Nair, S., Shankarappa, S.A. and Menon, D. (2016). Electrospun polymeric core-sheath yarns as drug eluting surgical sutures. *ACS Applied Materials and Interfaces*, 8(11), 6925–6934.

75. Bae, S., DiBalsi, M.J., Meilinger, N., Zhang, C., Beal, E., Korneva, G., Brown, R.O., Kornev, K. and Lee, J. (2018). Heparin-eluting electrospun nanofiber yarns for antithrombotic vascular sutures. *ACS Applied Materials and Interfaces*, 10(10), 8426–8435.

76. Padmakumar, S., Paul-Prasanth, B., Pavithran, K., Vijaykumar, D.K., Rajanbabu, A., Sivanarayanan, T.B., Kadakia, E., Amiji, M., Nair, S. and Menon, D. (2019). Long-term drug delivery using implantable electrospun woven polymeric nanotextiles. *Nanomedicine: Nanotechnology, Biology, and Medicine*, 15(1), 274–284.

77. Sun, G., Sun, L., Xie, H. and Liu, J. (2016). Electrospinning of nanofibers for energy applications. *Nanomaterials*, 6, 129.

78. Rahbar, R.S., Maleki, H. and Kalantari, B. (2016). Fabrication of electrospun nanofibre yarn based on nylon 6/microencapsulated phase change materials. *Journal of Experimental Nanoscience*, 11, 1402–1415.

79. Gao, H., Asheghali, D., Yadavalli, N.S., Pham, M., Nguyen, T., Minko, S. and Sharma, S. (2020). Fabrication of core-sheath nanoyarn via touchspinning and its application in wearable piezoelectric nanogenerator. *Journal of the Textile Institute*, 111, 906–915.

80. Ma, L., Zhou, M., Wu, R., Patil, A., Gong, H., Zhu, S., Wang, T., Zhang, Y., Shen, S., Dong, K., Yang, L., Wang, J., Guo, W. and Wang, Z. (2020). Continuous and scalable manufacture of hybridized nano-micro triboelectric yarns for energy harvesting and signal sensing. *ACS Nano*, 14(4), 4716–4726.

81. Liu, P., Wu, S., Zhang, Y., Zhang, H. and Qin, X. (2016). A fast response ammonia sensor based on coaxial PPy–PAN nanofiber yarn. *Nanomaterials*, 6(7), 121.

82. Wu, S., Liu, P., Zhang, Y., Zhang, H. and Qin, X. (2017). Flexible and conductive nanofiber-structured single yarn sensor for smart wearable devices. *Sensors and Actuators B-Chemical*, 252, 697–705.

83. Zhou, Y., He, J., Wang, H., Qi, K., Nan, N., You, X., Shao, W., Wang, L., Ding, B. and Cui, S. (2017). Highly sensitive, self-powered and wearable electronic skin based on pressure-sensitive nanofiber woven fabric sensor. *Scientific Reports*, 7, article no. 12949.

84. Nan, N., He, J., Xiaolu, Y., Xianqiang, S., Yuman, Z., Qi, K., Wei-li, S., Fan, L., Yan-yan, C. and Ding, B. (2019). A Stretchable, highly sensitive, and multimodal mechanical fabric sensor based on electrospun conductive nanofiber yarn for wearable electronics. *Advanced Materials and Technologies*, 4, 1800338.

85. Qi, K., Wang, H., You, X., Tao, X., Li, M., Zhou, Y., Zhang, Y., He, J., Shao, W. and Cui, S. (2019). Core-sheath nanofiber yarn for textile pressure sensor with high pressure sensitivity and spatial tactile acuity. *Journal of Colloid and Interface Science*, 561, 93–103.

86. Weerasinghe, V.T., Dissanayake, D., Perera, W.T., Tissera, N., Wijesena, R. and Wanasekara, N. (2020). All-organic, conductive and biodegradable yarns from core–shell nanofibers through electrospinning. *RSC Advances*, 10, 32875–32884.
87. Wang, Y., Yokota, T. and Someya, T. (2021). Electrospun nanofiber-based soft electronics. *NPG Asia Materials*, 13, 1–22.
88. Tsai, Chen-Chih, (2013). Electrospun nanofiber yarns for nanofluidic applications. All dissertations, 1609. https://tigerprints.clemson.edu/all_dissertations/1609.
89. Strong J.T., Misch C.E., Bidez M.W. and Nalluri P. (1998). Functional surface area: Thread-form parameter optimization for implant body design. *Compendium of Continuing Education in Dentistry*, 19(special), 4–9.
90. Rai, A.A., Stojanovska, E., Fidan, G., Yetgin, E., Polat, Y., Kılıç, A., Demir, A. and Yilmaz, Ş. (2020). Structure and performance of electroblown PVDF-based nanofibrous electret filters. *Polymer Engineering and Science*, 60, 1186–1193.
91. Peng, H., Liu, Y. and Ramakrishna, S. (2017). Recent development of centrifugal electrospinning. *Journal of Applied Polymer Science*, 134, 44578.
92. Kancheva, M., Toncheva, A., Manolova, N. and Rashkov, I. (2014). Advanced centrifugal electrospinning setup. *Materials Letters*, 136, 150–152.
93. Chen, H., Li, X., Li, N. and Yang, B. (2017). Electrostatic-assisted centrifugal spinning for continuous collection of submicron fibers. *Textile Research Journal*, 87, 2349–2357.
94. Chen, R., Li, Y. and He, J. (2014). Mini-review on Bubbfil spinning process for mass-production of nanofibers. *Materia-rio De Janeiro*, 19, 325–343.
95. Ren, Z. and He, J. (2011). Single polymeric bubble for the preparation of multiple micro/nano fibers. *Journal of Applied Polymer Science*, 119, 1161–1165.
96. Dou, H., Kuang, K., Li, Y., Fan, W., Shen, Y., Liu, H. and He, J. (2021). Effect of solution concentrations on the structure and properties of nanofibrous yarns by blown bubble-spinning. *Thermal Science*, 25(3), 2155–2160.
97. Stylios, G.K. (2014). Electrocarding technology: The manufacturing of nanoyarns for mass production of nanofabrics. *International Journal*

of Clothing Science and Technology, 26. https://doi.org/10.1108/IJCST-06-2014-0075

98. Zhou, F. and Gong, R. (2008). Manufacturing technologies of polymeric nanofibres and nanofibre yarns. *Polymer International*, 57, 837–845.
99. Sawhney, A., Condon, B., Singh, K.V., Pang, S., Li, G. and Hui, D. (2008). Modern applications of nanotechnology in textiles. *Textile Research Journal*, 78, 731–739.
100. Song, J., Kim, M. and Lee, H. (2020). Recent advances on nanofiber fabrications: Unconventional state-of-the-art spinning techniques. *Polymers*, 12, 1386.

Chapter 4

Drug-Loaded Nanofibers: Production Techniques and Release Behaviors

Hülya Kesici Güler and Funda Cengiz Çallioğlu
Süleyman Demirel University, Engineering Faculty,
Textile Engineering Department, Isparta, Turkey
hulyakesici@sdu.edu.tr

Electrospun nanofibers have gained importance in recent decades, especially in biomedical applications. Specifically, nanofibers can be used as drug carriers for controlled release and drug delivery aims. There are various methods to produce drug-loaded nanofibers, such as blend, suspension, emulsion, coaxial electrospinning, etc. These drug-loaded nanofiber production techniques are different from each other according to the way the drugs are located in the fibers. The different positioning of drugs into the fibers has a great effect on their release behavior. The drug release behavior of nanofibers is one of the most important factors that determine the suitable application area in terms of release time, amount, etc. Drug-loaded electrospun nanofibers can be used as drug delivery devices, either implantable or non-implantable, such as transdermal, oral, or topical.

Handbook of Nanofibers and Nanocomposites: Characteristics, Synthesis, and Applications in Textiles
Edited by Mohd Yusuf and Aminoddin Haji

ISBN 978-981-4968-77-5 (Hardcover), 978-1-003-43274-6 (eBook)
www.jennystanford.com

In this chapter, the production methods of drug-loaded nanofibers and comparative analysis of these approaches in respect of release behaviors, time, and amount were explained in detail along with current developments.

4.1 Introduction

Drug delivery techniques have been thoroughly researched for many years. Recent years have seen a significant increase in interest in electrospun fiber-based drug delivery methods among the scientific community as a result of their unique properties [1]. Electrospun fiber mats have high porosity and can have their properties optimized by adjusting the processing parameters. The fibers have a high surface area-to-volume ratio, which provides a large surface area for high loading capacity [2]. To achieve optimal therapeutic outcomes, an ideal formulation should be biocompatible, cost-effective, and capable of delivering drugs at predetermined rates [3].

The polymer carrier chosen, and the exact electrospinning technique selected can be used to control drug release [4]. In general, as mentioned below, there are six ways of loading bioactive compounds into fibers during electrospinning as described in the literature: surface modification, blending directly with the polymer solution, suspension, coaxial, emulsion, and microcapsulation into the electrospun nanofibers. It is possible to use the surface modification method for rapid drug release since the drug molecules are placed on the surface of the drug delivery vehicle in this method. Blend and coaxial electrospinning approaches have been shown to have some drawbacks. For instance, the integrated drugs or bioactive proteins in the blended electrospun fibers exhibit a burst release phenomenon at the first release point. The tendency of drugs or bioactive proteins deposited on the surface of the fibers under the intense electrostatic force experienced during the electrospinning process may be the cause of this burst release phenomenon. Therefore, core-shell nanofibers are preferable because they inhibit the burst release phenomenon. The coaxial electrospinning method is the most widely used method to produce core-shell nanofibers. However, coaxial electrospinning has its own set of challenges, such as the necessity for specialized equipment and the need to fine-tune the conditions to achieve the desired outcomes [5, 6]. The suspension prepared during the suspension electrospinning approach ensures that the medication molecules are distributed uniformly throughout the nanofiber during the manufacturing process. Emulsion electrospinning is an easier and more efficient approach than these two methods. As a consequence, this technique has gained popularity in recent years. Emulsion electrospinning provides various advantages, such as the incorporation of bioactive compounds into nanofibers as a core-sheath, prolonged release of bioactive molecules from nanofibers, and suitability for hydrophilic components sensitive to organic solvents [7]. Microcapsule-loaded nanofibers, which are made through a combination of micro and nanotechnology, can be used where long-term release is required.

It is changing the way drugs are administered to patients because nanofibers, which are being used as drug delivery systems, have the capability of regulating drug delivery. Drugs are delivered alone or in combination with excipients in single doses in traditional dosage formulations. They have a high bioavailability rate, but also a high clearance rate. Drugs that need to be taken quickly, such as anti-inflammatory, pain-reducing, and antibiotics, can be given in this dosage mode. However, they must be taken at regular intervals for long-term and chronic use. Drug delivery systems extend this time period and provide patients with more suitable dosing. Nonetheless, combining with a drug delivery system enables the elimination of drug degradation, the elimination of drug systemic effects, and the targeting of drug release to desired locations [8, 9]. Drug encapsulation aims to alleviate these major drawbacks.

Therefore, in this chapter, various electrospinning approaches that are used to load drugs onto the nanofibers were analyzed deeply, such as surface modification, blend, suspension, coaxial, and emulsion electrospinning. Moreover, the addition of drug-loaded microcapsules to the nanofibers is a new generation method to provide long-term drug release applications.

4.2 Production Techniques

Electrospinning allows for a wide range of materials for drug delivery via various techniques. Through the use of various electrospinning techniques, such as surface modification, blend, coaxial, suspension, emulsion electrospinning, and microcapsule-loaded nanofibers, there is a significant potential for drug incorporation into nanofiber structures.

4.2.1 Surface Modification

Surface modification with intended biomolecules is another technology that can be used to combine bio-functionality and nanofibers effectively. In this process, the therapeutic agent and drugs are bound to the nanofiber surfaces. The active ingredient release period of a product can be controlled by designing single or multi-layer sandwich structures coated with the active ingredient, resulting in a more consistent product [10]. A schematic representation of the surface modification process is given in Fig. 4.1.

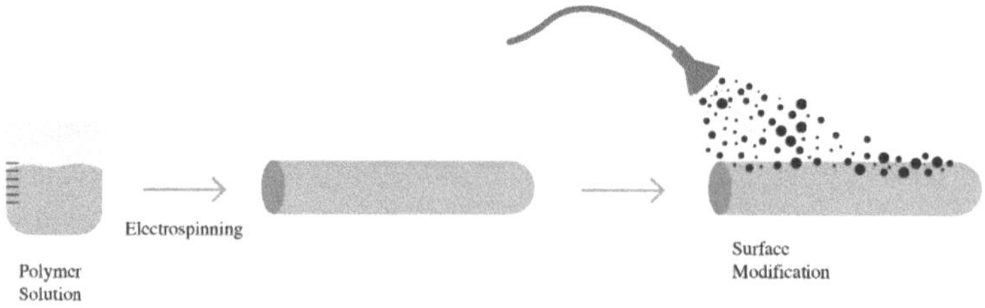

Figure 4.1 Schematic representation of surface modification.

Adding functionalities to nanofiber surfaces through physical adsorption is a simple and basic method of preparing nanofibers for drug delivery. A high amount of drugs can be loaded onto nanofibers because of the nanofibers' large specific surface area. However, the adsorbed molecules are frequently released too quickly, and the surface modification systems are probably more suited for short-term drug delivery. The surface characteristics of electrospun fibers can be controlled by the adsorption/desorption ratio. The drug adsorption and desorption behaviors were connected to drug characteristics such as kPa value (A drug's pKa is the ionization level at which it is 50% ionized), lipophilicity, and molecule size and solvent parameters. This situation demonstrates the importance of complicated adjustment of various factors in order to achieve a correct balance between drug adsorption and desorption rate and optimum delivery period for molecules bound on the surface of fibers [8].

Electrospun fibers were modified to incorporate different types of drugs. For example, ciprofloxacin was combined with PAN electrospun nanofibers via the surface adsorption method [11]. Another study is about Fe_3O_4/PAN and composite electrospun nanofibers produced via a two-step method such as electrospinning and a subsequent solvothermal process. Researchers indicated that nanofibrous composite structures have simple preparation procedures, high adsorption capacity, good regenerability, low cost, and an environmentally benign nature [12, 13].

However, the major disadvantage of this method is that drugs that must be endocytosed or those that must interact with the cell nucleus are unable to be immobilized in this technique [13].

In addition to the immobilization of drug molecules on the fibers themselves, surface modification of nanofibers with a wide range of chemical compounds can be used to modulate drug release profiles from drug-blended nanofibers. For instance, fluorination of the nanofiber surfaces resulted in a regulated rate of drug release by adding hydrophobic functional groups onto the fiber surface [10, 14].

4.2.2 Blend Electrospinning

Blend electrospinning involves incorporating a drug with a polymeric solution prior to the electrospinning process. The drug is dissolved or dispersed in the polymeric matrix, resulting in an encapsulated drug with a delayed release profile. In polymeric matrixes, the release of substances is governed either by desorption/diffusion or by dissolution/erosion of the polymeric matrix. The diffusion of the drug through the polymer membrane is responsible for the release of the drug from nonbiodegradable polymers. However, a factor related to system dissolution or erosion must be considered for biodegradable polymers. As a result, the release is typically controlled by either diffusion or erosion.

Blend electrospinning is thought to be easier to do than core/sheath electrospinning approaches. However, biomolecules may lose their bioactivity due to conformational changes in the organic solution environment. It is still important to examine the effects of various electrospinning methods on functional biomolecules with complex structures in more depth. Furthermore, earlier research has shown that blend and core/sheath nanofibers have different release characteristics [15]. Figure 4.2 depicts a schematic representation of blend electrospinning.

Figure 4.2 Schematic representation of blend electrospinning.

The combination of drugs with nanofibers and polymer solutions is still the most important of all the advanced technologies of drug loading and integration into nanofibers. Blend electrospinning, which is also called one-phase electrospinning, is used to disperse or dissolve the drug in the polymer solution, resulting in an encapsulated drug. Many researchers have blended and incorporated hydrophilic and hydrophobic polymers using various polymer blends in order to achieve sustained release. Their findings have shown that adding hydrophilic polymers can significantly improve drug-loading efficiency. As a result, reducing drug burst release has a positive effect on the polymer and the hydrophilic drug, but may impede the drug molecules' ability to transfer to the nanofibers' surface. On the other hand, it has been shown that polymer modification and copolymerization improve the hydrophilic properties of polymeric carriers [13].

In most studies about drug-loaded nanofibrous matrixes, direct electrospinning of drug/polymer blends is preferred [16–20]. The idea of achieving continuous drug release rather than the more instantaneous or burst release typical of post-spinning adsorptive drug loading has piqued people's interest in employing drug/polymer blends for electrospinning [21].

Antibiotics, cytostatic agents, and anti-inflammatory drugs have all been tested using blend electrospinning. Regrettably, the majority of biocompatible polymers, such as PCL, PLA, and PU, are soluble in organic solvents. Because of the harsh environment, it is difficult to use blend electrospinning to encapsulate molecules. As a possible consequence, biomolecules delivered via this method are often less bioactive. In addition to this, polar polymers have a high degree of solubility and short release times. One possibility is that cross-linking can be used to regulate the stability of nanofibers [22–25].

The solubility of a drug in a polymer solution is expected to affect its distribution in electrospun fibers and the degree of burst release. Moreover, for electrospinning drug/polymer solution blends, drug activities are likely to be affected by electrical charge and mechanical stress during electrospinning and exposure to organic solvents required for the dissolution of many polymers. In addition, the rate of polymer degradation can influence drug release from electrospun fibers. On the other hand, the presence of the second drug or additive may change the release kinetics of the first when producing fibrous matrixes electrospun from polymer blends containing two drugs or additives [21].

Covalent coupling of drugs to the polymer before electrospinning, as well as coating of drug-loaded fibers after electrospinning, were used to try to better control the release of drugs from electrospun matrixes. This method enabled prolonged drug release from electrospun matrixes under mild aqueous conditions [21, 26].

Drug combination with nanofibers and polymer solutions continues to be one of the most advanced ways of drug loading and integration into nanofibers. The drug is distributed or dissolved in the polymer solution in order to prepare the encapsulated drug, which is done by the blend electrospinning process, also known as the one-phase electrospinning technique, for this purpose. Many researchers have investigated the blending and incorporation of hydrophilic and hydrophobic polymers into various polymer blends with the aim of reaching sustained release [15, 27, 28]. Their findings have revealed that the addition of hydrophilic polymers can significantly improve drug-loading efficiency while simultaneously decreasing the burst release of drugs [13]. The positive contact between the polymer and the hydrophilic drug may help to reduce the tendency of drug molecules to transfer to the surface of the nanofibers [29].

4.2.3 Suspension Electrospinning

Suspension is defined as a mixture of liquid and solid particles uniformly dispersed in the liquid. Solid particles of suspension size range from 0.5 nm to 100 μm. The primary benefit of suspensions is that they dissolve active compounds that are insoluble in any solvent. In addition to the liquid and solid phases, other substances can be used to prepare the suspensions more stable. Wetting agents, flocculating agents, viscosity enhancers, buffers and pH adjusters, antimicrobials, antioxidants, and surfactants are examples of these. A schematic representation of suspension electrospinning is shown in Fig. 4.3.

Figure 4.3 Schematic representation of suspension electrospinning.

From the literature, it has been known that drugs can be encapsulated into nanofibers using various techniques while producing nanofibers that can release drugs. Suspension electrospinning is one of them, and this novel field holds great promise for green electrospinning, which is so important for sustainability and the environment [7]. Also, there is no need for an additional apparatus or a change in the production mechanism while using the suspension electrospinning method. Using this process, proteins, drugs, and various bioactive compounds can be added to the polymer solutions and deposited into the nanofibers. According to this method, an active substance insoluble in the polymer solution is suspended with the appropriate surfactant. The purpose of the suspension preparation here is to ensure that the active ingredient is dispersed as homogeneously as possible in the polymer solution. The suspension electrospinning method is very similar to the blending method. The most important aim of suspension is to suspend the drug-active substance that is insoluble in the polymer solution and to delay sinking the drug into the solution during the spinning process.

When the literature was examined, it was seen that the studies on the suspension electrospinning method were very new and very limited. In these studies carried out with the suspension electrospinning method; various polymers such as poly(L-lactide-co-glycolide) [30, 31], poly(3-hydroxybutyric acid) (PHB) [32], poly(hexamethylene adipate)-PEO block copolymers (PHA-b-PEO) [33], methoxy-poly (ethylene glycol) (PEG) [34], chitosan/poly(vinyl alcohol)/poly(ε-caprolactone) [35], polyethylene glycol/(PEG)–polyethylene oxide (PEO) [36] were used.

Naveen et al. (2010) carried out a study on nanofiber production by the suspension electrospinning method. In the study, poly-3-hydroxybutyrate was used due to its nontoxic properties, and kanamycin sulfate, a hydrophilic antibiotic, was used as a drug. In this research, the morphology, antimicrobial properties, drug release profile, cell viability, and cell morphology of the produced nanofibers were investigated. As a result, a high inhibition diameter was obtained against *Staphylococcus aureus* bacteria in the produced nanofibers. It was determined that as the polymer concentration increased, the surface tension, viscosity, and average fiber diameter also increased. Finally, it has been observed that cells multiply uniformly on the nanofiber surface [32].

In another study, Han et al. (2012) investigated nanofiber production by the emulsion and suspension electrospinning methods and also the release behavior of water-soluble bioactive substances from nanofibers. In the study, mixtures of poly(L-lactic-co-glycolic acid) (PLGA) and polyethylene glycol-b-poly(L-lactic-co-caprolactone) (PELCL) were used as polymers, and chitosan hydrogel was used as a carrier. Researchers examined the structure and morphology, hydrophilicity, mechanical properties, release behavior, and cyto-biocompatibility of the fibers obtained. As a result of the study, it was determined that the average fiber diameters of the obtained nanofibers were between 150 nm and 1380 nm, and all nanofiber surfaces showed good hydrophilicity, mechanical properties, and cyto-biocompatibility. Finally, it was determined that poly (ethylene glycol)-b-poly (L-lactide-co-caprolactone) nanofiber was released faster than poly (L-lactide-co-glycolide) nanofiber [30].

It is thought that the suspension electrospinning approach will be more attractive for researchers to various new electrospun nanofibers applications.

4.2.4 Coaxial Electrospinning

Coaxial electrospinning is a well-known method to generate drug-releasing nanofibers. Coaxial electrospinning is based on the co-spinning of two polymeric liquids at the same time. The general system for core/shell electrospinning uses two needles positioned coaxially. The spinneret permits the creation of composite polymeric droplets by pumping the inner (core) liquid via the internal needle and delivering the shell material through the outside needle. The spinneret permits the creation of composite polymeric droplets by pumping the inner (core) liquid via the internal needle and delivering the shell material through the outside needle [8]. A schematic presentation of coaxial electrospinning is given in Fig. 4.4.

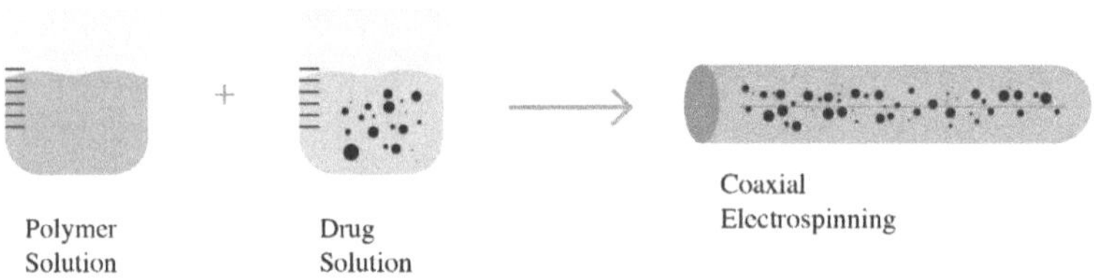

Figure 4.4 Schematic representation of coaxial nanofibers.

The processing conditions utilized for the core/shell method add numerous requirements to the conventional electrospinning process. The most significant factor is that the shell polymer is based on electrospinnable solutions as the molecular weight, concentration, and entanglement of polymeric chains must be appropriate to form stable fiber jets. As an alternative, a non-spinning liquid could be used as a core solution. In order to make core-shell fibers, the interfacial tensions between the two liquid phases must be high enough to allow the drawing of the core liquid while avoiding the mixing of the two polymeric solutions [37].

It is possible to produce composite functionalized nanofibers with a distinct core/shell structure using coaxial electrospinning. Some of the most significant advantages of coaxial electrospinning include the ability to form core-shell nanofibers from miscible and immiscible polymers, the high loading capacity of bioactive molecules, the extended-release of these molecules from the fibers, and a less harsh process that makes it possible to deliver susceptible compounds to the body. The biggest advantage of coaxial electrospinning is the ability to create electrospun fibrous meshes from materials that are either non-spinnable or only partly spinnable. In most cases, a well-electrospinnable polymer is used in the shell, allowing fibers to be formed from less electrospinnable solutions in the core [8].

It has been demonstrated that coaxial electrospinning may be utilized to deliver a wide range of drugs with excellent success, such as anti-inflammatory and analgesic drugs, antibacterial drugs, and anticancer drugs.

Using coaxial electrospinning is the preferred method for these compounds since it extends their release time. Coaxial electrospinning was shown to have a prolonged release time and less burst compared to blend electrospinning. The use of coaxial electrospinning allows for the release of biomolecules from hydrophilic matrixes to occur at a slower rate. For blend electrospinning, the release was faster at all time points compared to the other types of spinning. Electrospinning with a core-shell structure, on the other hand, enables the development of systems with long-term release behaviors. Diffusion coefficients between core and shell layers, as well as the drug type and degradation of core-shell materials, all have a significant impact on the drug release rate.

4.2.5 Emulsion Electrospinning

Emulsion electrospinning is another new approach to producing drug-loaded nanofibrous surfaces. An emulsion, as defined by the literature, is a dispersion of two or more immiscible liquids in between each other. There are two phases: continuous and dispersed in emulsion electrospinning. The basic rule is that droplets (in the dispersed phase) are spread uniformly throughout the continuous phase [7]. The preparation of emulsion nanofibers and the structure of emulsion electrospun nanofibers are given in Fig. 4.5.

The most important key factor is the solution preparation method for emulsion electrospinning. Utilizing emulsion electrospinning as a dispersed phase, hydrophilic and hydrophobic drugs can be incorporated into nanofibers. Furthermore, hydrophilic and hydrophobic compounds can be incorporated into nanofibers by utilizing emulsion electrospinning as a dispersed phase and aqueous or lyophilic solutions as a continuous phase. Also, both oil-in-water and water-in-oil emulsions can be used for producing core-sheath nanofibers via the electrospinning method [6, 38]. Hence, we may conclude that functional nanofiber production can be achieved by preparing various polymer/solvent/drug systems using the emulsion electrospinning approach.

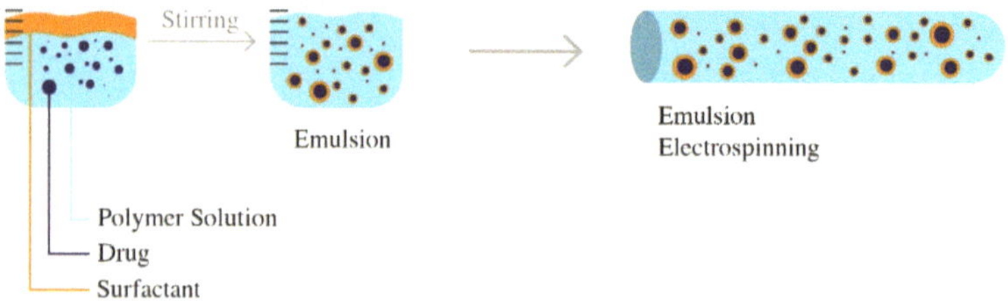

Figure 4.5 Preparation of emulsion nanofiber.

A new method known as emulsion electrospinning has the potential to solve many of the drawbacks of blend and coaxial electrospinning. The technology is a hybrid of the two processes, utilizing an emulsification technique. Electrospinning of emulsions is based on the production of a core-shell structure using a single nozzle. Unlike coaxial spinning, electrospinning of stable emulsions of two or more fluids provides core-shell fiber organization. In contrast to blend electrospinning, the emulsion consists of two or more separate phases that are not immiscible during the electrospinning process. Consequently, the electrospun solution is composed of two liquid phases, such as the continuous phase, which is formed into a shell of fibers, and the droplet phase, which is responsible for creating fiber cores. Immiscible phases are stabilized with surfactants [8].

In general, two types of emulsions are utilized in the electrospinning process: water-in-oil (o/w) and oil-in-water (w/o) emulsions. Water-in-oil emulsions are composed of a lipophilic continuous phase and a hydrophilic droplet phase. Using this kind of emulsion, you can encapsulate molecules that are either polar or hydrophilic. Surfactants of low-value HLB (Hydrophilic-Lipophilic Balance) stabilize this type of emulsion, such as Span60 and Span80

[39, 40]. Emulsions of oil-in-water (o/w) are the other type of emulsion. In this type of emulsion, a hydrophilic solution forms the continuous phase, while a lipophilic solution forms the droplet phase. Surfactants with high HLB values like Tween 20 and Tween 80 help stabilize the emulsion [41, 42].

As well as emulsion preparation, several properties are vital for the electrospinning process to be successful. A polymer solution as a continuous phase with electrospinnability must be used to create the emulsion. This means that the solution must have sufficient conductivity, and an appropriate polymer concentration, and that the polymer must have the necessary high molecular weight and surface tension to achieve the desired results. Inversely, droplet phase features affect the fiber's internal structure. During the emulsion electrospinning process, a continuous core is produced along the fiber axis when the viscosity of the droplet phase is higher than the viscosity of the continuous phase. The core phase, on the other hand, breaks up into unconnected droplets if its viscosity is lower than that of the shell phase, producing a noncontinuous morphology for the core phase [34, 43].

To begin with, Sanders et al. 2003 referred to this technology as two-phase electrospinning rather than emulsion electrospinning in the literature. The fundamentals of emulsion electrospinning are reinforced by two-phase electrospinning [44].

Liao et al. (2008) obtained nanofibers by using PLGA/chloroform: DMF solution, Span80 surfactant, and Rhodamine B/pure water solution to simulate the drug to be used. As a result, the desired hydrophilicity and release behavior of the nanofibers obtained by emulsion formation during solution preparation was obtained, considering the hydrophilicity and desired release behavior in tissue scaffolds [45].

Yan et al. (2009) encapsulated bovine serum albumin (BSA) and RhodaminB into nanofibers produced from poly (L-lactide-co-caprolactone) (PLLACL) polymer by the emulsion electrospinning method. The BSA/pure water solution, that is, the water phase, was prepared, and the emulsion was prepared by adding dropwise to the oil phase. When the *in vitro* release studies are examined, while the release of BSA did not differ in the samples containing BSA in the first 96 hours, they observed that after 96 hours there was more release from containing the double active ingredient than containing only

BSA. As a result, when the release behavior of five different samples was examined, it was determined that the nanofibers obtained by forming emulsions showed the desired and controllable release performance [46].

Xiaoqiang Li et al. (2010) encapsulated human nerve growth factor (NGF) and BSA into biodegradable poly (L-lactide-co-caprolactone) (PLACL) nanofibers by the emulsion electrospinning method. They then studied the protein release from these nanofibers. They found that the immediate release is very low and generally stable in the emulsion electrospinning method compared to the blend electrospinning method. Moreover, it was determined that the Span80 surfactant used to form the emulsion was on the surface of the nanofibers as a result of the hydrophilicity test [39].

Hu et al. (2015) produced nanofiber surfaces containing poly (lactide-co-glycolide) (PLGA) based gelatin-reinforced Cefradine (hydrophobic) and 5-fluorouracil (hydrophilic) drugs. They used both blend and emulsion electrospinning methods during production. According to the results obtained, the reason for this is that Span80, which is used as a surfactant while forming the emulsion, is a viscous liquid at room temperature and, because it has relatively high mobility in the production process polymer jet during nanofiber production, it moves to the surface of the fibers and remains there, and because the gelatin polymer contains more charge groups than PLGA, it becomes attached to the surface of the fibers. They explained it as forcing the fibers to move to the surface under electrical force while the surfaces have the desired mechanical properties. Finally, in the release studies, they determined that the emulsion electrospun nanofibers made a stable and sustainable release by having a low burst release compared to the nanofibers produced according to the blend method. As a result, they stated that the obtained PLGA/gelatin emulsion electrospun nanofiber surfaces can be used not only as tissue scaffolds but also as desired and controllable drug release vehicles [47].

Tian et al. (2012) used vascular endothelial growth factor (VEGF), BSA, and dextran (DEX) on poly (L-lactic acid-co-ε-caprolactone) (PLCL) nanofibers in their study encapsulated. For this purpose, both emulsion and blend PLCL polymer solutions were prepared with appropriate formulas. According to the obtained characterization results, they determined that the water phase, namely VEGF, is located in the center of the nanofibers and core-shell type nanofibers are produced. Finally, according to the results obtained from *in vitro* release studies, they divided the VEGF release process from PLCL nanofibers into two stages: the first 24 hours and the last 648 hours. In the first stage, the hydrophilic proteins on the surface of the nanofibers were directly released into the buffer solution, PBS, while in the second stage, they observed that controlled and slow release occurred by drug diffusion as a result of drug diffusion from the core of the nanofibers and degradation of the polymer. VEGF release from nanofibrous scaffolds continued for a total of 28 days. As a result, the researchers stated that the emulsion electrospinning method is a suitable and convincing approach for different types of tissue regeneration with controlled drug and protein release from nanofibers [48].

4.2.6 Microcapsule Loaded Nanofibers

A microcapsule is a substance that is generated when a solid, liquid, or gas-phase substance is encapsulated within a shell. Microcapsulation techniques have several advantages, including the ability to protect substances from unsuitable or adverse environmental conditions, reduce evaporation and diffusion rates, extend their useful life by preventing disintegrating reactions like oxidation and dehydration, achieve controlled and targeted drug release, and enhance durability properties such as fluidity, dispersibility, and improved solubility.

Despite the fact that a variety of procedures are utilized in the production of drug-loaded microcapsules, it is observable that the emulsion solvent evaporation approach is the most prevalent [49–52]. The emulsion/solvent evaporation method is the most suitable method for the microencapsulation of water-insoluble drugs. In addition, the fact that drug-loaded microcapsules do not leak or crack after production makes this method even more advantageous. During production, the production is carried out in a simpler way since the solvent can be evaporated quickly by itself, by temperature, or by mixing. That is, there is no need for an extra addition to the system [53, 54].

Nanofibers have nanosized fiber diameters, high porosity, a small and controllable open pore structure, very high loading capacity, and a high specific surface area, which are used to create new generation textile surfaces and to prepare drug delivery systems.

In the literature, there are a lot of studies about drug-loaded microcapsules such as furosemide [55], indomethacin [56], ketoprofen [57], ibuprofen [58], imatinib [59], and pamidronate-disodium [60]. But the drug-loaded microcapsule-loaded nanofibrous structure is a new approach for preparing smart drug delivery systems, which exist only in very limited literature. For instance, paracetamol-loaded microcapsules were produced via the emulsion/solvent evaporation method. These microcapsules were added to PVA polymer solutions at various concentrations. Then, nanofibers were produced with electrospinning under the optimum process parameters. Researchers demonstrated paracetamol-loaded microcapsule-loaded nanofibers with SEM pictures [61]. Another study is about St. John's Worth oil, which is frequently used in the medical field and is loaded as a core material with PVA and Eudragit RS 100 polymers as a shell material for producing microcapsules. The aim of this study is to develop new generations of drug delivery devices using micro and nanotechnology in combination. They clearly observed from SEM images that the number of microcapsules in the nanoweb increased as the concentration of microcapsules increased [62].

In Fig. 4.6, schematic representations of drug-loaded microcapsules and microcapsule-loaded nanofibers were shown.

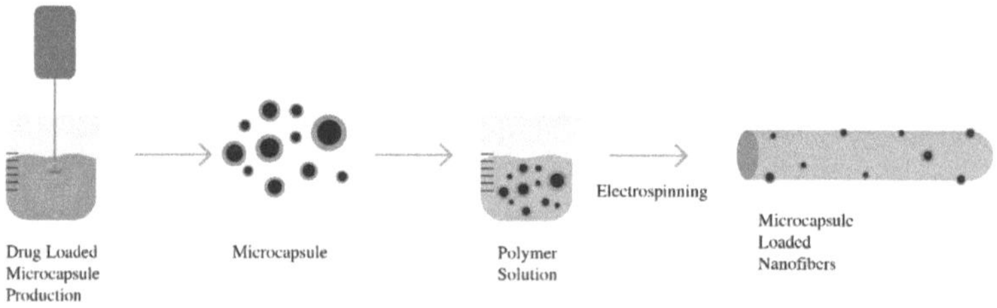

Figure 4.6 Schematic representation of microcapsule-loaded nanofibers.

The initial step in this procedure, as shown in Fig. 4.6, is to microencapsulate the drug using the appropriate method and polymer. The microcapsules are then blended with the suitable polymer solution under the proper conditions. Finally, nanofibers under the optimum process parameters must be produced from the obtained polymer solution containing the microcapsules. Figure 4.7 shows the SEM image of microcapsules loaded nanofibers produced with this approach at different magnifications.

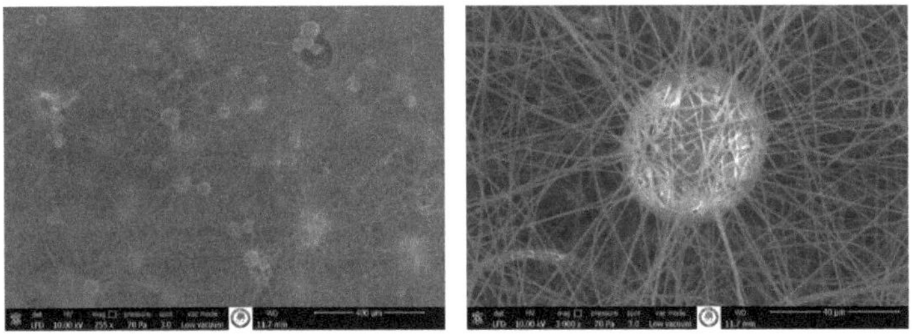

(a) (b)

Figure 4.7 SEM images of microcapsule loaded nanofiber with different magnifications: (a) 255x, (b) 3000x.

This approach includes two aims. The first is to utilize microcapsule technology to protect highly sensitive biomolecules from harmful environmental conditions. Because organic and hazardous solvents used in the preparation of polymer solutions can harm some biomolecules. The second step is to encapsulate the microcapsules in nanofibers and convert them to the nanofibrous surface.

This approach allows for parameter-based optimization of drug release behavior. The first is the concentration of the polymer that forms the microcapsule's shell, which controls the shell thickness and therefore the core material's release. The second is the concentration of core material applied in microcapsule production. In summary, the effect of the wall/core material ratio on the release is quite significant. The third parameter is the concentration of the microcapsule to be added to the polymer solution prepared for nanofiber production. Unlike these, other effective parameters include the characteristics of the polymer to be used as the shell material, the properties of the core material, and the fabrication method.

It is thought that this approach will receive more attention in the near future and has the potential to be used as a drug delivery system.

4.3 Conclusion

Nanotechnology is currently having a considerable impact in the fields of medical diagnostics, drugs, and biotechnology, among other fields. The advantages of using electrospinning technology to create nanofibers as drug delivery systems have not yet been completely realized. There are a variety of techniques for incorporating biomolecules and drugs into nanofiber materials that include surface modification, blend, suspension, coaxial, emulsion, and microcapsule-loaded nanofibers. Due to their simplicity in design, these nanofibers were considered carriers for drug delivery that improved pharmacokinetic parameters such as blood-residence times, delivering an effective method to a goal. In this chapter, the production methods of nanofibrous drug delivery systems produced by electrospinning, drug-loaded nanofibers' release behaviors, and application areas determination according to their release behaviors are discussed and compared with the literature.

References

1. Bhardwaj, N., and Kundu, S.C. (2010). Electrospinning: A fascinating fiber fabrication technique. *Biotechnology Advances*, **28**, 325–347. 10.1016/j.biotechadv.2010.01.004.
2. Li, H., Liu, K., Sang, Q., Williams, G.R., Wu, J., Wang, H., Wu, J., and Zhu, L.-M. (2017). A thermosensitive drug delivery system prepared by blend electrospinning. *Colloids and Surfaces B: Biointerfaces*, **159**, 277–283. 10.1016/j.colsurfb.2017.07.058.
3. Qiu, Y., and Park, K. (2001). Environment-sensitive hydrogels for drug delivery. *Advanced Drug Delivery Reviews*, **53**, 321–339. 10.1016/s0169-409x(01)00203-4.
4. Sill, T.J., and Von Recum, H.A. (2008). Electrospinning: Applications in drug delivery and tissue engineering. *Biomaterials*, **29**, 1989–2006. 10.1016/j.biomaterials.2008.01.011.
5. Briggs, T., and Arinzeh, T.L. (2014). Examining the formulation of emulsion electrospinning for improving the release of bioactive proteins from electrospun fibers. *Journal of Biomedical Materials Research Part A*, **102**, 674–684. 10.1002/jbm.a.34730.
6. Hu, J., Wei, J., Liu, W., and Chen, Y. (2013). Preparation and characterization of electrospun PLGA/gelatin nanofibers as a drug

delivery system by emulsion electrospinning. *Journal of Biomaterials Science, Polymer Edition*, **24**, 972–985.

7. Agarwal, S., and Greiner, A. (2011). On the way to clean and safe electrospinning—green electrospinning: Emulsion and suspension electrospinning. *Polymers for Advanced Technologies*, **22**, 372–378.
8. Buzgo, M., Mickova, A., Rampichova, M., and Doupnik, M. (2018). Blend electrospinning, coaxial electrospinning, and emulsion electrospinning techniques. In: *Core-shell Nanostructures for Drug Delivery and Theranostics*, Elsevier, pp. 325–347.
9. Yang, W.-W., and Pierstorff, E. (2012). Reservoir-based polymer drug delivery systems. *Journal of Laboratory Automation*, **17**, 50–58. 10.1177/2211068211428189.
10. Zamani, M., Prabhakaran, M.P., and Ramakrishna, S. (2013). Advances in drug delivery via electrospun and electrosprayed nanomaterials. *International Journal of Nanomedicine*, **8**, 2997. 10.2147/IJN.S43575.
11. Li, X., Chen, S., Fan, X., Quan, X., Tan, F., Zhang, Y., and Gao, J. (2015). Adsorption of ciprofloxacin, bisphenol and 2-chlorophenol on electrospun carbon nanofibers: In comparison with powder activated carbon. *Journal of Colloid and Interface Science*, **447**, 120–127. 10.1016/j.jcis.2015.01.042.
12. Liu, Q., Zhong, L.-B., Zhao, Q.-B., Frear, C., and Zheng, Y.-M. (2015). Synthesis of Fe3O4/polyacrylonitrile composite electrospun nanofiber mat for effective adsorption of tetracycline. *ACS Applied Materials and Interfaces*, **7**, 14573–14583. 10.1021/acsami.5b04598.
13. Mohammadian, F., and Eatemadi, A. (2017). Drug loading and delivery using nanofibers scaffolds. *Artificial Cells, Nanomedicine, and Biotechnology*, **45**, 881–888.
14. Im, J.S., Yun, J., Lim, Y.-M., Kim, H.-I., and Lee, Y.-S. (2010). Fluorination of electrospun hydrogel fibers for a controlled release drug delivery system. *Acta Biomaterialia*, **6**, 102–109. 10.1016/j.actbio.2009.06.017.
15. Ji, W., Yang, F., Van den Beucken, J.J., Bian, Z., Fan, M., Chen, Z., and Jansen, J.A. (2010). Fibrous scaffolds loaded with protein prepared by blend or coaxial electrospinning. *Acta Biomaterialia*, **6**, 4199–4207. 10.1016/j.actbio.2010.05.025.
16. Hong, Y., Fujimoto, K., Hashizume, R., Guan, J., Stankus, J.J., Tobita, K., and Wagner, W.R. (2008). Generating elastic, biodegradable polyurethane/poly (lactide-co-glycolide) fibrous sheets with controlled antibiotic release via two-stream electrospinning. *Biomacromolecules*, **9**, 1200–1207. 10.1021/bm701201w.

17. Um-i-Zahra, S., Shen, X.X., Li, H., and Zhu, L. (2014). Study of sustained release drug-loaded nanofibers of cellulose acetate and ethyl cellulose polymer blends prepared by electrospinning and their in-vitro drug release profiles. *Journal of Polymer Research*, **21**, 1–12.

18. Valarezo, E., Tammaro, L., Malagón, O., González, S., Armijos, C., and Vittoria, V. (2015). Fabrication and characterization of poly (lactic acid)/poly (ε-caprolactone) blend electrospun fibers loaded with amoxicillin for tunable delivering. *Journal of Nanoscience and Nanotechnology*, **15**, 4706–4712.

19. Vashisth, P., Sharma, M., Nikhil, K., Singh, H., Panwar, R., Pruthi, P.A., and Pruthi, V. (2015). Antiproliferative activity of ferulic acid-encapsulated electrospun PLGA/PEO nanofibers against MCF-7 human breast carcinoma cells. **3** *Biotech*, **5**, 303–315. 10.1007/s13205-014-0229-6.

20. Yang, G., Wang, J., Li, L., Ding, S., and Zhou, S. (2014). Electrospun micelles/drug-loaded nanofibers for time-programmed multi-agent release. *Macromolecular Bioscience*, **14**, 965–976. 10.1002/mabi.201300575.

21. Meinel, A.J., Germershaus, O., Luhmann, T., Merkle, H.P., and Meinel, L. (2012). Electrospun matrices for localized drug delivery: Current technologies and selected biomedical applications. *European Journal of Pharmaceutics and Biopharmaceutics*, **81**, 1–13. 10.1016/j.ejpb.2012.01.016.

22. Mickova, A., Buzgo, M., Benada, O., Rampichova, M., Fisar, Z., Filova, E., Tesarova, M., Lukas, D., and Amler, E. (2012). Core/shell nanofibers with embedded liposomes as a drug delivery system. *Biomacromolecules*, **13**, 952–962.

23. Song, D.W., Kim, S.H., Kim, H.H., Lee, K.H., Ki, C.S., and Park, Y.H. (2016). Multi-biofunction of antimicrobial peptide-immobilized silk fibroin nanofiber membrane: Implications for wound healing. *Acta Biomaterialia*, **39**, 146–155. 10.1016/j.actbio.2016.05.008.

24. Wang, X., Yue, T., and Lee, T.-c. (2015). Development of Pleurocidin-poly (vinyl alcohol) electrospun antimicrobial nanofibers to retain antimicrobial activity in food system application. *Food Control*, **54**, 150–157.

25. Zehetmeyer, G., Meira, S.M.M., Scheibel, J.M., da Silva, C.d.B., Rodembusch, F.S., Brandelli, A., and Soares, R.M.D. (2017). Biodegradable and antimicrobial films based on poly (butylene adipate-co-terephthalate) electrospun fibers. *Polymer Bulletin*, **74**, 3243–3268.

26. Kenawy, E.-R., Abdel-Hay, F.I., El-Newehy, M.H., and Wnek, G.E. (2007). Controlled release of ketoprofen from electrospun poly (vinyl alcohol) nanofibers. *Materials Science and Engineering: A*, **459**, 390–396.

27. Islam, M.S., and Karim, M.R. (2010). Fabrication and characterization of poly (vinyl alcohol)/alginate blend nanofibers by electrospinning method. *Colloids and Surfaces A: Physicochemical and Engineering Aspects*, **366**, 135–140.

28. Li, H., Sang, Q., Wu, J., Williams, G.R., Wang, H., Niu, S., Wu, J., and Zhu, L.-M. (2018). Dual-responsive drug delivery systems prepared by blend electrospinning. *International Journal of Pharmaceutics*, **543**, 1–7. 10.1016/j.ijpharm.2018.03.009.

29. Zamani, M., Morshed, M., Varshosaz, J., and Jannesari, M. (2010). Controlled release of metronidazole benzoate from poly ε-caprolactone electrospun nanofibers for periodontal diseases. *European Journal of Pharmaceutics and Biopharmaceutics*, **75**, 179–185. 10.1016/j.ejpb.2010.02.002.

30. Han, F., Zhang, H., Zhao, J., Zhao, Y., and Yuan, X. (2012). In situ encapsulation of hydrogel in ultrafine fibers by suspension electrospinning. *Polymer Engineering and Science*, **52**, 2695–2704.

31. Han, F., Zhang, H., Zhao, J., Zhao, Y., and Yuan, X. (2013). Diverse release behaviors of water-soluble bioactive substances from fibrous membranes prepared by emulsion and suspension electrospinning. *Journal of Biomaterials Science, Polymer Edition*, **24**, 1244–1259. 10.1080/09205063.2012.746510.

32. Naveen, N., Kumar, R., Balaji, S., Uma, T., Natrajan, T., and Sehgal, P. (2010). Synthesis of nonwoven nanofibers by electrospinning–a promising biomaterial for tissue engineering and drug delivery. *Advanced Engineering Materials*, **12**, B380–B387.

33. Sun, J., Bubel, K., Chen, F., Kissel, T., Agarwal, S., and Greiner, A. (2010). Nanofibers by green electrospinning of aqueous suspensions of biodegradable block copolyesters for applications in medicine, pharmacy and agriculture. *Macromolecular Rapid Communications*, **31**, 2077–2083. 10.1002/marc.201000379.

34. Xu, X., Yang, L., Xu, X., Wang, X., Chen, X., Liang, Q., Zeng, J., and Jing, X. (2005). Ultrafine medicated fibers electrospun from W/O emulsions. *Journal of Controlled Release*, **108**, 33–42. 10.1016/j.jconrel.2005.07.021.

35. Zargarian, S., and Hadadiasl, V. (2010). A nanofibrous composite scaffold of PCL/hydroxyapatite-chitosan/PVA prepared by electrospinning. *Iranian Polymer Journal*, 2010, **19**(6), 457–468

36. Zhang, H., and Edirisinghe, M. (2006). Electrospinning zirconia fiber from a suspension. *Journal of the American Ceramic Society* **89**, 1870–1875.

37. Vysloužilová, L., Buzgo, M., Pokorný, P., Chvojka, J., Míčková, A., Rampichová, M., Kula, J., Pejchar, K., Bílek, M., and Lukáš, D. (2017). Needleless coaxial electrospinning: A novel approach to mass production of coaxial nanofibers. *International Journal of Pharmaceutics*, **516**, 293–300. 10.1016/j.ijpharm.2016.11.034.

38. Gordon, V., Marom, G., and Magdassi, S. (2015). Formation of hydrophilic nanofibers from nanoemulsions through electrospinning. *International Journal of Pharmaceutics*, **478**, 172–179.

39. Li, X., Su, Y., Liu, S., Tan, L., Mo, X., and Ramakrishna, S. (2010). Encapsulation of proteins in poly (L-lactide-co-caprolactone) fibers by emulsion electrospinning. *Colloids and Surfaces B: Biointerfaces*, **75**, 418–424. 10.1016/j.colsurfb.2009.09.014.

40. Tadros, T. (2006). Principles of emulsion stabilization with special reference to polymeric surfactants. *Journal of Cosmetic Science*, **57**, 153–169.

41. Arecchi, A., Mannino, S., and Weiss, J. (2010). Electrospinning of poly (vinyl alcohol) nanofibers loaded with hexadecane nanodroplets. *Journal of Food Science*, **75**, N80-N88. 10.1111/j.1750-3841.2010.01680.x.

42. Camerlo, A., Bühlmann-Popa, A.-M., Vebert-Nardin, C., Rossi, R.M., and Fortunato, G. (2014). Environmentally controlled emulsion electrospinning for the encapsulation of temperature-sensitive compounds. *Journal of Materials Science*, **49**, 8154–8162.

43. Angeles, M., Cheng, H.L., and Velankar, S.S. (2008). Emulsion electrospinning: Composite fibers from drop breakup during electrospinning. *Polymers for Advanced Technologies*, **19**, 728–733.

44. Sanders, E.H., Kloefkorn, R., Bowlin, G.L., Simpson, D.G., and Wnek, G.E. (2003). Two-phase electrospinning from a single electrified jet: Microencapsulation of aqueous reservoirs in poly (ethylene-co-vinyl acetate) fibers. *Macromolecules*, **36**, 3803–3805.

45. Liao, Y., Zhang, L., Gao, Y., Zhu, Z.-T., and Fong, H. (2008). Preparation, characterization, and encapsulation/release studies of a composite nanofiber mat electrospun from an emulsion containing poly (lactic-co-glycolic acid). *Polymer*, **49**, 5294–5299. 10.1016/j.polymer.2008.09.045.

46. Yan, S., Xiaoqiang, L., Shuiping, L., Xiumei, M., and Ramakrishna, S. (2009). Controlled release of dual drugs from emulsion electrospun nanofibrous mats. *Colloids and Surfaces B: Biointerfaces*, **73**, 376–381. 10.1016/j.colsurfb.2009.06.009.

47. Hu, J., Prabhakaran, M.P., Ding, X., and Ramakrishna, S. (2015). Emulsion electrospinning of polycaprolactone: Influence of surfactant type towards the scaffold properties. *Journal of Biomaterials Science, Polymer Edition*, **26**, 57–75.

48. Tian, L., Prabhakaran, M.P., Ding, X., Kai, D., and Ramakrishna, S. (2012). Emulsion electrospun vascular endothelial growth factor encapsulated poly (L-lactic acid-co-ε-caprolactone) nanofibers for sustained release in cardiac tissue engineering. *Journal of Materials Science*, **47**, 3272–3281.

49. Amasya, G., Badilli, U., Aksu, B., and Tarimci, N. (2016). Quality by design case study 1: Design of 5-fluorouracil loaded lipid nanoparticles by the W/O/W double emulsion—Solvent evaporation method. *European Journal of Pharmaceutical Sciences*, **84**, 92–102. 10.1016/j.ejps.2016.01.003.

50. Behera, B., Sahoo, S., Dhal, S., Barik, B., and Gupta, B. (2008). Characterization of glipizide-loaded polymethacrylate microspheres prepared by an emulsion solvent evaporation method. *Tropical Journal of Pharmaceutical Research*, **7**, 879–885.

51. Kim, B., Hwang, S., Park, J., and Park, H.J. (2002). Preparation and characterization of drug-loaded polymethacrylate microspheres by an emulsion solvent evaporation method. *Journal of Microencapsulation*, **19**, 811–822.

52. M. Obeidat, W., and Price, J.C. (2006). Preparation and evaluation of Eudragit S 100 microspheres as pH-sensitive release preparations for piroxicam and theophylline using the emulsion-solvent evaporation method. *Journal of Microencapsulation*, **23**, 195–202.

53. Heiskanen, H., Denifl, P., Pitkänen, P., and Hurme, M. (2012). Effect of concentration and temperature on the properties of the microspheres prepared using an emulsion–solvent extraction process. *Advanced Powder Technology* **23**, 779–786.

54. Hong, Y., Gao, C., Shi, Y., and Shen, J. (2005). Preparation of porous polylactide microspheres by emulsion-solvent evaporation based on solution induced phase separation. *Polymers for Advanced Technologies*, **16**, 622–627.

55. Ai, H., Jones, S.A., de Villiers, M.M., and Lvov, Y.M. (2003). Nano-encapsulation of furosemide microcrystals for controlled drug release. *Journal of Controlled Release*, **86**, 59–68.

56. Liu, H., Finn, N., and Yates, M. (2005). Encapsulation and sustained release of a model drug, indomethacin, using CO_2-based microencapsulation. *Langmuir*, **21**, 379–385.

57. Palmieri, G., Bonacucina, G., Martino, P.D., and Martelli, S. (2002). Gastro-resistant microspheres containing ketoprofen. *Journal of Microencapsulation*, **19**, 111–119. 10.1080/02652040110065477.

58. Perumal, D. (2001). Microencapsulation of ibuprofen and Eudragit® RS 100 by the emulsion solvent diffusion technique. *International Journal of Pharmaceutics*, **218**, 1–11.

59. Ramazani, F., Chen, W., Van Nostrum, C., Storm, G., Kiessling, F., Lammers, T., Hennink, W.E., and Kok, R. (2015). Formulation and characterization of microspheres loaded with imatinib for sustained delivery. *International Journal of Pharmaceutics*, **482**, 123–130. 10.1016/j.ijpharm.2015.01.043.

60. Weidenauer, U., Bodmer, D., and Kissel, T. (2003). Microencapsulation of hydrophilic drug substances using biodegradable polyesters. Part I: Evaluation of different techniques for the encapsulation of pamidronate di-sodium salt. *Journal of Microencapsulation*, **20**, 509–524.

61. Mol, İ.Y., Geysoğlu, M., Kesici Güler, H., and Cengiz Çallıoğlu, F. (2021). Paracetamol drug loaded microcapsule based nanofiber production. In: *TEXTEH Proceedings*, **2021**, 200–207.

62. Kesici Güler, H., Cengiz Çallıoğlu, F., Mol, İ.Y., and Geysoğlu, M. (2021). Electrospinning of St. John's wort oil loaded microcapsules based PVA nanofibers. In: *TEXTEH Proceedings*, **2021**, 171–177.

Chapter 5

Textile Applications of Nanofibers and Nanocomposites

Shumaila Kiran, Shahid Adeel, Shazia Abrar, Sarosh Iqbal, Saba Naz, and Nimra Amin
Department of Applied Chemistry, Government College University, Faisalabad 38000, Pakistan
shahidadeel@gcuf.edu.pk

Nanotechnology is revolutionizing the global community by introducing day-by-day sustainable products. The proper inclusion of nanoparticles in the polymeric material can enhance its performance to a much extent. These things are well-designated as nanofilled polymer composites. The use of multifunctional textiles in computing systems, consumer products, electronics, health, medical sciences, sports, power energy storage, and other applications has increased the demand for nanomaterials in the textile sector. Hence, four major textile industry domains that are using nanotechnology, nanofinishing, nanofilms, nanostructured materials, and composite materials, are explored in this chapter. In addition, various

Handbook of Nanofibers and Nanocomposites: Characteristics, Synthesis, and Applications in Textiles
Edited by Mohd Yusuf and Aminoddin Haji

ISBN 978-981-4968-77-5 (Hardcover), 978-1-003-43274-6 (eBook)
www.jennystanford.com

functional fabrics using nanoparticles are briefly discussed. New desirable characteristics like flame-retarding nature, enhanced biodegradability, antimicrobial characteristics, etc. can be induced in the nanostructured composites prepared for different walks of life, particularly for textiles.

5.1 Introduction

In 1959, Richard Feynman presented the nanotechnology concept in his work which he delivered at the California Institute of Technology's American Physical Symposium. Until 1974, when Norio Taniguchi came up with the concept of "nanotechnology," Feynman's ideas were unreported [1, 2]. One nanometer is one-billionth of a meter since the term "nano" denotes one-billionth or 10^{-9}. There are now a variety of manufacturing technologies available that can design and produce nanostructures to the desired composition, shape, morphology, size, and crystalline nature. "Top-down" and "bottom-up" are the two most common manufacturing processes. Nanotechnology is already contributing to significant prospects in designing new materials with numerous advantages for use in various applications. Because the nanosized atoms inside nanoparticles are properly arranged, the material characteristics undergo significant changes [3].

Researchers can currently synthesize nanosized materials including carbon nanotubes, nanoclays, carbon nanotubes, nanofibers, and teflon that are stronger, lighter, have higher chemical stability, and have better control over the spectrum of light [4, 5]. A better knowledge of the characteristics of nanoparticles will create opportunities for the sustainable development of novel materials in the future years, with the potential to improve living quality [6]. Nanomaterials are increasingly becoming globalized, progressing as products, and being utilized in a wide range of advanced technology applications and electronics, consisting of a wide range of customer items [7]. In the future, it is expected that the design and formation of nanoparticles using a unique blend of fiber will enhance the demanding duration [8]. In addition, synthetic nanoparticles are being widely investigated by scientists along with corporations, in order to improve existing functionality in goods as well as develop new ones [9]. Despite such advancements in nanomaterial technology, literature on the potential consequences of nanomaterials on the environment and human health has been insufficient till now [10]. Exposure to artificial nanomaterials has been associated with a variety of health effects, therefore nanosafety is a serious issue [11, 12]. Green nanoscience has been proposed as a solution to reduce the potential environmental concerns and hazards to human health associated with the manufacturing and use of nanomaterials, as well as to create the replacement of common objects with the most environmental-friendly nanomaterials [13, 14].

5.2 Nanomaterials in Textile Industry

Currently, the textile sector is a major nanotechnology handler, and a huge amount of nanotextiles are available in the retailing shops [15–17]. Nanotextiles are standard fabrics that have nanoparticles added to them. The ability to resist fire (flame retardancy), self-deterging, water resistance, the ability to resist water (dirt repellency), antibacterial properties, and UV radiation protection, are just a few of the characteristics exhibited in these modern fabrics [18, 19]. Nanocomposites and nanofinishings are improving the increase of applications for fabric in multiple sectors [20, 21]. The use of nanoparticles and a nanostructure-based protective coating has exhibited significant potential in new functional and high-performance textiles [22–24]. A comprehensive study shows that the modification and characterization of textiles, using nanochemistry has revolutionized the globe [25]. The sustainable application of nanotechnology in textiles for its various eco-friendly functions has been displayed in Fig. 5.1. Nanomaterials have a greater potential for generating different functions in fabrics due to their nanoscale dimensions and the larger surface-area-to-volume ratio [26]. Various nanomaterials are typically used for textile applications:

(1) Carbon-based nanomaterials: carbon nanotubes, graphene, and carbon nanofibers
(2) Inorganic nanoparticles: metal oxide, metal, and nanoclay
(3) Nanoparticles with a core-shell
(4) Nanomaterials composites

(5) Nanomaterial hybrids
(6) Nanomaterials made up of polymers

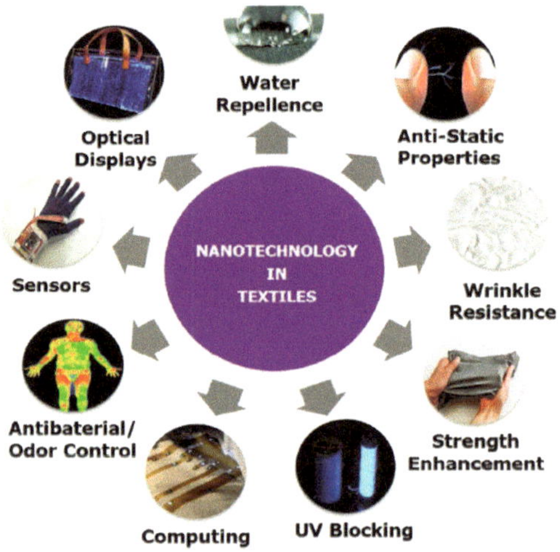

Figure 5.1 Utilization of nanotechnology in various walks of life.

5.3 Nanotechnology Applications in the Textile Industry

Nanomaterials and nanotechnology are applied in four major sectors of the textile industry are discussed below. Following their use, the materials are used in several fields (Table 5.1).

5.3.1 Nanofinishing

Nanofinishing is the technique of applying an ultrafine scattering of nanoparticles to a textile item to improve its properties [33, 34].

It offers several advantages over traditional finishing, the most important of which are

- In general, a smaller amount of nanomaterials is used in nanofinishing compared to bulk materials to achieve effective outcomes.
- Because of the greater surface availability of nanoparticles, as well as their homogeneous distribution in textile material, these nanofinishings are more durable.

- Nanofinishing may be able to generate certain behaviors that are difficult to achieve with ordinary finishes [35, 36].

Table 5.1 Fabric functionalization using nanoparticles

S. No.	Nanomaterial	Role	Citation
1.	Nanoclays	Flame retardance, dynamic component maintenance etc.	[27]
2.	Aluminum oxide	Abrasion resistance, resistance to fire	[28]
3.	Silicon dioxide	Water repellence, abrasion resistance, reinforcement enhanced the dyeability, repellent to dirt	[29]
4.	Zinc oxide	Self-cleaning, stiffness, resistance to abrasion, UV protection, and antimicrobial properties	[30]
5.	Titanium dioxide	UV protection, water repellence, self-cleaning, dirt repellence,	[31]
6.	Silver	Antibacterial property, conductive to electricity	[32]

5.3.2 Nanocoating

Traditional coatings are a very thick finish, making them extremely durable against imperfections such as chips, corrosions, scratches, or damage from road debris. It also allows for customization in texture and colored finishes. Traditional coatings, on the other hand, have several drawbacks, including

(i) lack of strength,
(ii) less variability, and
(iii) insufficient adherence [37].

The use of nanocoatings might overcome the aforementioned issues with regular coatings [38, 39]. The coating of nanomaterials on textiles does not affect their breathability [40].

Electrospinning, force spinning, self-assembly, island-in-sea, and melt blowing are some of the processes utilized to create nanofibers (bicomponent nanofiber) [41, 42]. Electrospinning is regarded as the most efficient of these approaches since it is inexpensive, increases porosity, greater production rate, and capacity to adjust nanofiber structure and thickness [43]. Initially, a piezo-resistive carbon nanotube nanofiber sensing yarn using carbon nanotubes was used. Figure 5.2 proposes the manufacturing process using a simple electrospinning approach [44]. In addition, the nanofibers are distinctive in that they have a great potential for use as active materials in face masks to protect patients from illnesses like the coronavirus [45].

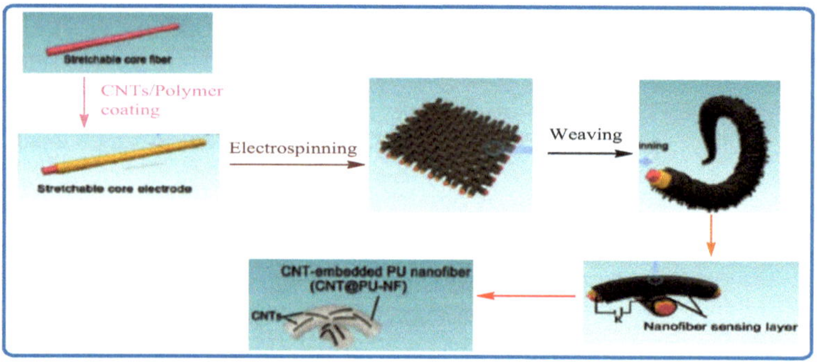

Figure 5.2 The production of carbon nanotube-incorporated nanofiber sensing yarn is illustrated in this diagram.

5.3.3 Nanocomposites

A multiphase solid substance is known as a nanocomposite in which the reinforcing phase has at least one dimension that is on the nanoscale [46]. Nanocomposites are distributed in matrixes made of polymers in polymer-based nanocomposites. Coatings based on polymer nanocomposite and fibers possess great potential in the manufacture of useful and high-performance textiles. Using a simple one-pot technique, new nanocomposite materials have been created. Silver nanoparticles and organic phosphates, and diphosphate malonate (DPHM) are used to manufacture nanocomposites [47].

To increase their fire retardancy and antibacterial qualities, several textile materials have been treated with the nanocomposites produced. The flame retardancy of the treated fabrics was found to be significantly increased [48].

5.4 Functional Nanomaterial Textiles

Textiles with nanomaterial coatings have a variety of functional qualities. Their properties reveal that the methods used for their synthesis or modification are sustainable, economic, energy-efficient, and highly biodegradable for the globe. These functional characteristics are as follows:

5.4.1 UV Protective Textiles

Ultraviolet radiation is now penetrating the earth's surface, causing harm to the worldwide community. People who are exposed to more UV radiation are more likely to experience a range of harmful disorders, particularly skin cancer. As a result, protection against UV radiation is a significant property for clothes and fabrics [49]. When metal-mediated nanomaterials like magnetite nanoparticles, zinc oxide nanoparticles, and titanium dioxide nanoparticles are used, they can reduce the severity of UV radiation effects [50–52]. Due to their excellent functionality, such additives are currently recommended over those that have been the subject of a recent study. The properties of nanosized UV absorbers, such as particle size, surface chemistry, physicochemical characteristics, thermodynamic properties, and crystal structure, have all had a serious influence on their ultraviolet inhibiting property [53, 54]. The strong UV protection and significant antibacterial efficiency of this ZnO-coated cotton textile have enhanced the characteristics. Researchers used polyurethane-based MnO_2-$FeTiO_3$ nanocomposites to produce ultraviolet light and lightning cotton fabric. The MnO_2-$FeTiO_3$-treated cotton textiles exhibited long-term ultraviolet shielding and improved toughness when tested using the confined oxygen index. While examined employing the low oxygen index, the MnO_2-$FeTiO_3$ coated cotton textiles revealed long-term UV protection and improved fire resistance. Furthermore, despite 10 water-laundering cycles, the coated cotton fabric preserved its qualities, making it a smart, sustainable, and durable fabric for use in protective garments [49]. As a result, nanotechnology has improved the protective textile industry by increasing its functional behavior.

5.4.2 Flame-Retardant Textiles

Flame-resistant fabrics are one of the most current issues of research in recent decades. According to recent studies, using nanoparticles may strengthen the textile polymers' flame-retardant characteristics and thermal stability [55–58]. The basic mechanism involved in the formation of a protective film is enhanced by char propagation and free radical trapping [59]. However, the extent to which things have improved is influenced by nanoparticle structure, morphology, and surface compatibility [60].

Researchers studied that UV protection and flame-retardant finishes were obtained by cross-linking the cotton with nanomaterials. They discovered that composite coated with polypyrrole, zinc oxide, and carbon nanotubes may exhibit good functional properties such as flame-retardant fabrics [61]. A previous study explored flame-retardant formulations using standard thermochromic flame retardant and nanosilica was produced and then applied to the cotton to explore the synergic effect of nanosilica on fire resistance and thermal stability. Their findings show that introducing nanosilica into a typical thermochromic fire-resistant solution can increase the flammability of cotton fabric [62].

5.4.3 Water- and Oil-Repellent Textiles

Textiles' permeability to water and grease is another valuable feature [63]. Improved nanocoatings or nanofinishings now satisfy the majority of oil and water-repellent textile items. Two basic methodologies are frequently employed in the production of self-cleaning fabrics: (i) photocatalytic activity and (ii) the lotus effect [64, 65].

Surface modification of textiles by nanocoating or nanofinishing, using ZnO nanorods, nanosilica, produces the lotus effect [66]. To manufacture self-cleaning textiles using the photocatalytic activity approach, nanosized ZnO or TiO_2 layers are used. Researchers proved that Au/TiO_2/ SiO_2 nanosol was a better photocatalyst than titanium dioxide nanosol, which has a better self-cleaning characteristic [67]. Several additional studies have used TiO_2 nanowire and titanium dioxide nanowire doped Ag-PVP and graphene/TiO_2 nanocomposites to exhibit the cotton fabric's photocatalytic self-cleaning properties [68]. Fluorine-free superhydrophobic cotton textiles comprised of superhydrophobic SiO_2 and photoactive titanium dioxide exhibited better self-cleaning photocatalytic activity [69, 70].

5.4.4 Antimicrobial Textiles

Textile made of cellulose fibers such as viscose, lyocell, cotton, and linen are more vulnerable to microbes [71]. Antimicrobial properties have recently been a requirement for all dyed fabrics, medical textiles, and home items as a result of increased awareness about hygiene and health. Metal oxides in various forms (such as ZnO, TiO_2, and CuO) and metal nanoparticles (NPs) such as titanium, zinc, silver, gold, silica, and copper nanoparticles have recently received great interest as potential antibacterial agents [72]. In comparison to a single nanomaterial, the use of antimicrobial nanocomposites in textiles revealed a good synergistic antibacterial activity. Cotton fibers coated with silver nanoparticles made from natural Chinese holly plant extracts have antibacterial characteristics, which are both cost-effective and environmentally beneficial [73]. Using a UV-Vis spectrophotometer, the application of Ag-NPs from Chinese Holly plant extracts was determined. Against gram-negative *Escherichia coli* bacteria, the antibacterial activities of these coated cotton fibers combined with Ag-NPs were tested. Using standard methods, they discovered that combining 1.5 percent to 4.5 percent of Chinese Holly leaf extracts resulted in better antibacterial activities. After multiple items of washing, the cotton fibers also showed excellent antibacterial activity, making them suitable for medicinal applications [74]. A comprehensive overview of improved antimicrobial characteristics of textiles coated with nanoparticles is given in Table 5.2.

Table 5.2 Textiles modified using different nanoparticles for antimicrobial characteristics

S. No.	Nanoparticles	Size	Fiber	Microorganisms	Results	Reference
1.	Silver	2.3 nm	Cotton	*E. coli* and *S. aureus*	99% reduction for *Staphylococcus aureus* and 92% reduction for *E. coli*	[75]
2.	Silver	60–100 nm	Cotton and rayon	*E. faecalis, S. aureus,* and *E. coli*	*E. faecalis* 97%, *S. aureus*-98%, and *E. coli*-100%	[76]
3.	Copper oxide	83 nm	Cotton	*K. pneumonaie, E. coli,* and *S. aureus*	Superior antimicrobial activity with antimicrobial durability of 93% subsequent to 50 washes.	[77]
4.	Titanium dioxide	50 nm	Cotton	*E. coli* and *S. aureus*	Greater than 95% decrease subsequent to 20 washes	[78]
5.	Titanium dioxide –graphene oxide	Less than 100 nm	Cellulose acetate fibers	*B. cereus* and *B. subtilis*	Greater than 95% reduction	[79]

5.4.5 Wrinkle-Resistant Fabrics

Cotton fabric is extremely vulnerable to crease formation during usage. Resin-based finishings are widely applied in standard methods to provide crease resistance to textile materials. The use of nanoparticles like silicon dioxide and titanium dioxide can overcome some of the drawbacks of conventional crease-resistant treatments [22, 80–83]. Scientists assessed the wrinkling behavior and wrinkle resistance ability of cotton textiles dyed with Direct Blue 2B in the absence and presence of titanium dioxide nanoparticles. Their findings revealed that direct dyeing improved the wrinkle resistance of cross-linked materials [84]. Researchers studied the padding process used to examine the wrinkle-resistant, advanced flame retardant, and the durable finish of linen with nano-additives. They reported that flax textiles treated with aluminum oxide nanoparticles had better flame-retardant and wrinkle-resistant qualities [85].

5.5 Future Prospects

The area of nanotechnology is still in its early stages, with multiple issues, challenges, and commercial potential. Modern science, particularly analytical chemistry, is bringing a wide range of techniques that might be used to monitor nanoparticles on textiles and in textile wastewater. Research must be conducted to prevent possible human health dangers for users, including adults, especially newborns and children.

5.6 Conclusions

The material at the nanoscale has gained significant attention in textile applications. These materials are promising for the functionalization of fibers and textiles. The nanotextiles applications are possible because of the characteristics of nanomaterials such as wear/tear resistance, antistatic properties, UV protection, antimicrobial property, flame retardancy, wrinkle-free, and water/oil/dirt repellency. Though, due to abrasion or on washing, such textiles discharge nanoparticles (NPS). The NPS gather in soil and aquatic bodies, making them harmful to the environment and human health. It is necessary to develop separation and characterization techniques for the identification and quantification of nanoparticles from soil and water. Eco-toxicological research must be done to determine the environmental deterioration and effect because of NPS on human health.

References

1. Saleem, H., Zaidi, S.J. (2020). Sustainable use of nanomaterials in textiles and their environmental impact, *Materials*, 13(22), 5134, 1–28.
2. Mirshafiee, V., Osborne, O.J. Sun, B., Xia, T. (2018). Safety concerns of industrial engineered nanomaterials. In: *Handbook of Nanomaterials for Industrial Applications* (Elsevier Amsterdam, the Netherlands), 1063–1072.
3. Yang, G., Park, S.J. (2019). Deformation of single crystals, polycrystalline materials, and thin films: A Review, *Materials*, 12, 2003, 1–18.
4. Khan, I., Saeed, K., Khan, I. (2019). Nanoparticles: Properties, applications and toxicities, *Arab. J. Chem.*, 12, 908–931.
5. Ehrmann, A., Nguyen, T.A., Tri, P.N. (2020). *Nanosensors and Nanodevices for Smart Multifunctional Textiles* (Elsevier Amsterdam, the Netherlands).
6. Ahmad, F., Abubshait, S.A., Abubshait, H.A. (2020). Untargeted metabolomics for Achilles heel of engineered nanomaterials' risk assessment, *Chemosphere*, 262, 128058.
7. Schulte, P.A., Kuempel, E.D., Drew, N.M. (2018). Characterizing risk assessments for the development of occupational exposure limits for engineered nanomaterials, *Regul. Toxicol. Pharmacol.*, 95, 207–219.
8. Verma, A., Arif, R., Jadoun, S. (2020). Synthesis, characterization, and application of modified textile nanomaterials. In: Shabbir, M., Ahmed, S., Sheikh, J.N. (eds.) *Frontiers of Textile Materials: Polymers, Nanomaterials, Enzymes, and Advanced Modification Techniques* (Scrivener Publishing LLC Beverly, MA, USA), 167–187.
9. Ahmed, S.F., Mofijur, M., Rafa, N., Chowdhury, A.T., Chowdhury, S., Nahrin, M., Ong, H.C. (2022). Green approaches in synthesising nanomaterials for environmental nanobioremediation: Technological advancements, applications, benefits and challenges, *Environ. Res.*, 204, 111967.
10. Kumar, R., Chauhan, M., Sharma, N., Chaudhary, G.R. (eds.) (2018). Toxic effects of nanomaterials on environment. In: *Environmental Toxicity of Nanomaterials* (CRC Press, Boca Raton, FL, USA), 1–20.

11. Johnston, L.J., Gonzalez-Rojano, N., Wilkinson, K.J., Xing, B. (2020). Key challenges for evaluation of the safety of engineered nanomaterials. *NanoImpact*, 18, 100219.
12. Karim, M.E. (2020). Functional nanomaterials: Selected occupational health and safety concerns. In: Hussain, C.M. (ed.), *Handbook of Functionalized Nanomaterials for Industrial Applications* (Elsevier Amsterdam, the Netherlands), 995–1006.
13. Iavicoli, I., Leso, V., Ricciardi, W., Hodson, L.L., Hoover, M.D. (2014). Opportunities and challenges of nanotechnology in the green economy, *Environ. Health*, 13, 78.
14. Bamoharram, F.F. (2011). Role of polyoxometalates as green compounds in recent developments of nanoscience. *Synth. React. Inorg. Met. Org. Nano-Met. Chem.*, 41, 893–922.
15. Jatoi, A.S., Khan, F.S.A., Mazari, S.A., Mubarak, N.M., Abro, R., Ahmed, J., Baloch, H., Sabzoi, N. (2021). Current applications of smart nanotextiles and future trends. In: Ehrman, A., Nguyen, T., Nguyen Tri, P. (eds.), *Nanosensors and Nanodevices for Smart Multifunctional Textiles* (Elsevier Amsterdam, the Netherlands), 343–365.
16. Darwesh, O.M., Ali, S.S., Matter, I.A., Elsamahy, T. (2021). Nanotextiles waste management: Controlling of release and remediation of wastes. In: Ehrman, A., Nguyen, T., Nguyen Tri, P. (eds.), *Nanosensors and Nanodevices for Smart Multifunctional Textiles* (Elsevier Amsterdam, the Netherlands), 267–286.
17. Schoden, F. (2021). Ecological and sustainable smart nanotextiles. In: Ehrman, A., Nguyen, T., Nguyen Tri, P. (eds.), *Nanosensors and Nanodevices for Smart Multifunctional Textiles* (Elsevier Amsterdam, the Netherlands), 287–320.
18. Xue, CH., Wu, Y., Guo, X.J., Liu, B.Y., Wang, H.D., Jia, S.T. (2020). Superhydrophobic, flame-retardant and conductive cotton fabrics via layer-by-layer assembly of carbon nanotubes for flexible sensing electronics, *Cellulose*, 27, 3455–3468.
19. Gadkari, R.R., Ali, S.W., Joshi, M., Rajendran, S., Das, A., Alagirusamy, R. (2020). Leveraging antibacterial efficacy of silver loaded chitosan nanoparticles on layer-by-layer self-assembled coated cotton fabric, *Int. J. Biol. Macromol.*, 162, 548–560.
20. Banerjee, B. (ed.) (2019). *Rubber Nanocomposites and Nanotextiles: Perspectives in Automobile Technologies* (Walter de Gruyter GmbH & Co KG., Berlin, Germany).
21. Jadoun, S., Verma, A., Arif, R. (2020). Modification of Textiles via Nanomaterials and Their Applications. In: Shabbir, M., Ahmed,

S., Sheikh, J.N. (eds.), *Frontiers of Textile Materials: Polymers, Nanomaterials, Enzymes, and Advanced Modification Techniques* (Scrivener Publishing LLC Beverly, MA, USA), 135–152.

22. Haque, M. (2019). Nanofabrics in the 21st century: A review, *Asian J. Nanosci. Mater.*, 2, 120–256.
23. Silva, I.O., Ladchumananandasivam, R., Nascimento, J.H.O., Silva, K.K.O., Oliveira, F.R., Souto, A.P., Felgueiras, H.P., Zille, A. (2019). Multifunctional chitosan/gold nanoparticles coatings for biomedical textiles, *J. Nanomater.*, 9, 1064.
24. Shabbir, M., Kaushik, M. (2020). Engineered nanomaterials: Scope in today's textile industry. In: Hussain, C.M. (ed.), *Handbook of Nanomaterials for Manufacturing Applications* (Elsevier Amsterdam, the Netherlands), 249–263.
25. Singh, M., Vajpayee, M., Ledwani, L. (2020). Eco-friendly surface modification and nanofinishing of textile polymers to enhance functionalisation. In: Ledwani, L., Sangwai, J.S. (eds.), *Nanotechnology for Energy and Environmental Engineering* (Springer, Cham, Switzerland), 529–559.
26. Yetisen, A.K., Qu, H., Manbachi, A., Butt, H., Dokmeci, M.R., Hinestroza, J.P., Yun, S.H. (2016). Nanotechnology in textiles, *ACS Nano*, 10, 3042–3068.
27. Gocek, I. (2019). Functionalization of textile materials with nanoclay incorporation for improved characteristics, *Politek. Derg.*, 22, 509–522.
28. Korkmaz, N., Alay., Aksoy, S. (2016). Enhancing the performance properties of ester-cross-linked cotton fabrics using Al_2O_3-NPs, *Text. Res. J.*, 86, 636–648.
29. Dogan, O., Dag, R. (2017). Application of nano-coating (SiO_2) on textile products, *J. Chem. Chem. Eng.*, 11, 82–85.
30. Verbic, A., Gorjanc, M., Simoncic, B. (2019). Zinc oxide for functional textile coatings: Recent advances, *Coatings*, 9, 550.
31. Abbas, M., Iftikhar, H., Malik, M.H., Nazir, A. (2019). Surface coatings of TiO_2 nanoparticles onto the designed fabrics for enhanced self-cleaning properties, *Coatings*, 8, 35.
32. Xu, Q., Xie, L., Diao, H., Li, F., Zhang, Y., Fu, F., Liu, X. (2017). Antibacterial cotton fabric with enhanced durability prepared using silver nanoparticles and carboxymethyl chitosan, *Carbohydr. Polym.*, 177, 187–193.

33. Ghosh, G., Sidpara, A., Bandyopadhyay, P.P. (2017). High efficiency chemical assisted nanofinishing of HVOF sprayed WC-Co coating, *Surf. Coat. Technol.*, 334, 204–214.

34. Radetic, M., Markovic, D. (2019). Nano-finishing of cellulose textile materials with copper and copper oxide nanoparticles, *Cellulose*, 26, 8971–8991.

35. Gokarneshan, N., Velumani, K. (2018). Significant trends in nano finishes for improvement of functional properties of fabrics. In: Yusuf, M. (ed.), *Handbook of Renewable Materials for Coloration and Finishing* (Scrivener Publishing LLC, Beverly, MA, USA), 387–434.

36. Ghosh, S., Smith, T., Rana, S., Goswami, P. (2020). *Nanofinishing of Textiles for Sportswear* (World Textile Information Network, Dublin, Ireland).

37. Nguyen-Tri, P., Nguyen, T.A., Carriere, P., Ngo., Xuan, C. (2018). Nanocomposite coatings: Preparation, characterization, properties, and applications, *Int. J. Corros.*, 2018, https://doi.org/10.1155/2018/4749501.

38. Joshi, M., Khanna, R., Shekhar, R., Jha, K. (2011). Chitosan nanocoating on cotton textile substrate using layer-by-layer self-assembly technique, *J. Appl. Polym. Sci.*, 119, 2793–2799.

39. Peng, L., Chen, W., Su, B, Yu, A., Jiang, X. (2019). CsxWO_3 nanosheet-coated cotton fabric with multiple functions: UV/NIR shielding and full-spectrum-responsive self-cleaning, *Appl. Surf. Sci.*, 475, 325–333.

40. Temesgen, A.G., Turşucular, Ö.F., Eren, R., Ulcay, Y. (2018). Novel applications of nanotechnology in modification of textile fabrics properties and apparel, *Int. J. Adv. Multidiscip. Res.*, 5, 49–58.

41. Qi, K., Zhou, Y., Ou, K., Dai, Y., You, X., Wang, H., He, J., Qin, X., Wang, R. (2020). Weavable and stretchable piezoresistive carbon nanotubes-embedded nanofiber sensing yarns for highly sensitive and multimodal wearable textile sensor, *Carbon*, 170, 464–476.

42. Nayak, R., Khandual, A. (2019). Chapter 18: Nanotextiles and recent developments. In: Patnaik, A., Patnaik, S. (eds.), *Fibres to Smart Textiles: Advances in Manufacturing, Technologies, and Applications* (CRC Press: Boca Raton, FL, USA).

43. Mamun, A., Sabantina, L., Klöcker, M., Heide, A., Blachowicz, T., Ehrmann, A. (2022). Electrospinning nanofiber mats with magnetite nanoparticles using various needle-based techniques, *Polymers*, 14(3), 533.

44. Zhang, Z., Wu, X., Kou, Z., Song, N., Nie, G., Wang, C., Mu, S. (2022). Rational design of electrospun nanofiber-typed electrocatalysts for water splitting: A review, *Chem. Eng. J.*, 428, 131133.

45. Tebyetekerwa, M., Xu, Z., Yang, S., Ramakrishna, S. (2020). Electrospun nanofibers-based face masks, *Adv. Fiber Mater.*, 2, 161–166.

46. Malhotra, B.D., Ali, M.A. (2018). *Nanomaterials for Biosensors: Fundamentals and Applications* (William Andrew: Norwich, NY, USA).

47. Attia, N.F., Morsy, M.S. (2016). Facile synthesis of novel nanocomposite as antibacterial and flame retardant material for textile fabrics, *Mater. Chem. Phys.*, 180, 364–372.

48. Shehabeldine, A.M., Hashem, A.H., Wassel, A.R., Hasanin, M. (2022). Antimicrobial and antiviral activities of durable cotton fabrics treated with nanocomposite based on zinc oxide nanoparticles, acyclovir, nanochitosan, and clove oil, *Appl. Biochem. Biotechnol.*, 194(2), 783–800.

49. Dhineshbabu, N.R., Bose, S. (2019). UV resistant and fire retardant properties in fabrics coated with polymer based nanocomposites derived from sustainable and natural resources for protective clothing application, *Compos. Part B Eng.*, 172, 555–563.

50. Attia, N.F., Moussa, M., Sheta, A.M., Taha, R., Gamal, H. (2017). Effect of different nanoparticles based coating on the performance of textile properties, *Prog. Org. Coat.*, 104, 72–80.

51. Fouda, A., Saad, E.L., Salem, S.S., Shaheen, T.I. (2018). In-Vitro cytotoxicity, antibacterial, and UV protection properties of the biosynthesized Zinc oxide nanoparticles for medical textile applications, *Microb. Pathog.*, 125, 252–261.

52. Sedighi, A., Montazer, M., Mazinani, S. (2018). Fabrication of electrically conductive superparamagnetic fabric with microwave attenuation, antibacterial properties and UV protection using PEDOT/magnetite nanoparticles, *Mater. Des.*, 160, 34–47.

53. Dhineshbabu, N.R., Bose, S. (2018). Smart textiles coated with eco-friendly UV-blocking nanoparticles derived from natural resources, *ACS Omega*, 3, 7454–7465.

54. Noorian, S.A., Hemmatinejad, N., Navarro, J.A. (2020). Ligand modified cellulose fabrics as support of zinc oxide nanoparticles for UV protection and antimicrobial activities, *Int. J. Biol. Macromol.*, 154, 1215–1226.

55. Ortelli, S., Malucelli, G., Blosi, M., Zanoni, I., Costa, A.L. (2019). NanoTiO_2@DNA complex: A novel eco, durable, fire retardant design strategy for cotton textiles, *J. Colloid Interface Sci.*, 546, 174–183.

56. Kundu, C.K., Song, L., Hu, Y. (2020). Nanoparticles based coatings for multifunctional Polyamide 66 textiles with improved flame retardancy and hydrophilicity, *J. Taiwan Inst. Chem. Eng.*, 112, 15–19.

57. Saleemi, S., Naveed, T., Riaz, T., Memon, H., Awan, J.A., Siyal, M.I., Xu, F., Bae, J. (2020). Surface functionalization of cotton and PC fabrics using SiO_2 and ZnO nanoparticles for durable flame retardant properties, *Coatings*, 10, 124.

58. Ali, S.W., Basak, S., Shukla, A. (2020). Nanoparticles: A potential alternative to classical fire retardants for textile substrates. In: Hussain, C.M. (ed.), *Handbook of Nanomaterials for Manufacturing Applications* (Elsevier Amsterdam, the Netherlands), 265–278.

59. Norouzi, M., Zare, Y., Kiany, P. (2015). Nanoparticles as effective flame retardants for natural and synthetic textile polymers: Application, mechanism, and optimization, *Polym. Rev.*, 55, 531–560.

60. Butola, B.S. (ed.) (2020). *Advances in Functional and Protective Textiles* (Woodhead Publishing: Cambridge, UK).

61. Yazhini, K.B., Prabu, H.G. (2015). Study on flame-retardant and UV-protection properties of cotton fabric functionalized with ppy–ZnO–CNT nanocomposite, *RSC Adv.*, 5, 49062–49069.

62. Fanglong, Z., Qun, X., Qianqian, F., Rangtong, L., Kejing, L. (2016). Influence of nano-silica on flame resistance behavior of intumescent flame retardant cellulosic textiles: Remarkable synergistic effect, *Surf. Coat. Technol.*, 294, 90–94.

63. Asif, A.K.M.A.H., Hasan, M.Z. (2018). Application of nanotechnology in modern textiles: A review, *Int. J. Curr. Eng. Technol.*, 8, 227–231.

64. Katiyar, P., Mishra, S., Srivastava, A., Prasad, N.E. (2020). Preparation of TiO_2–SiO_2 hybrid nanosols coated flame-retardant polyester fabric possessing dual contradictory characteristics of superhydrophobicity and self cleaning ability, *J. Nanosci. Nanotechnol.*, 20, 1780–1789.

65. Montazer, S.T.M. (2020). Denim fabric with flame retardant, hydrophilic and self-cleaning properties conferring by in-situ synthesis of silica nanoparticles, *Cellulose*, 27, 6643–6661.

66. Das, I., De, G. (2015). Zirconia based superhydrophobic coatings on cotton fabrics exhibiting excellent durability for versatile use, *Sci. Rep.*, 5, 18503.

67. Wang, R., Wang, X., Xin, J.H. (2010). Advanced visible-light-driven self-cleaning cotton by Au/TiO_2/SiO_2 photocatalysts, *ACS Appl. Mater. Interfaces*, 2, 82–85.

68. Hebeish, A.A., Abdelhady, M.M., Youssef, A.M. (2013). TiO_2 nanowire and TiO_2 nanowire doped Ag-PVP nanocomposite for antimicrobial and self-cleaning cotton textile, *Carbohydr. Polym.*, 91, 549–559.
69. Karimi, L., Yazdanshenas, M.E., Khajavi, R., Rashidi, A., Mirjalili, M. (2014). Using graphene/TiO2 nanocomposite as a new route for preparation of electroconductive, self-cleaning, antibacterial and antifungal cotton fabric without toxicity, *Cellulose*, 21, 3813–3827.
70. Xu, B., Ding, J., Feng, L., Ding, Y., Ge, F., Cai, Z. (2015). Self-cleaning cotton fabrics via combination of photocatalytic TiO_2 and superhydrophobic SiO_2, *Surf. Coat. Technol.*, 262, 70–76.
71. Bu, Y., Zhang, S., Cai, Y., Yang, Y., Ma, S., Huang, J., Yang, H., Ye, D., Zhou, Y., Xu, W. (2019). Fabrication of durable antibacterial and superhydrophobic textiles via in situ synthesis of silver nanoparticle on tannic acid-coated viscose textiles, *Cellulose*, 26, 2109–2122.
72. Attia, N.F., Moussa, M., Sheta, A.M., Taha, R., Gamal, H. (2017). Synthesis of effective multifunctional textile based on silica nanoparticles, *Prog. Org. Coat.*, 106, 41–49.
73. Ullah, N., Yasin, S., Abro, Z., Liu, L., Wei, Q. (2014). Mechanically robust and antimicrobial cotton fibers loaded with silver nanoparticles: Synthesized via Chinese holly plant leaves, *Int. J. Text. Sci.*, 3, 1–5.
74. Ouadil, B., Amadine, O., Essamlali, Y., Cherkaoui, O., Zahouily, M. (2019). A new route for the preparation of hydrophobic and antibacterial textiles fabrics using Ag-loaded graphene nanocomposite. *Colloids Surf. A Physicochem. Eng. Asp.*, 579, 123713.
75. Wu, Y., Yang, Y., Zhang, Z., Wang, Z., Zhao, Y., Sun, L. (2019). Fabrication of cotton fabrics with durable antibacterial activities finishing by Ag nanoparticles, *Text. Res. J.*, 89, 867–880.
76. Toh, H.S., Faure, R.L., Amin, L.B.M., Hay, C.Y.F., George, S. (2017). A light-assisted in situ embedment of silver nanoparticles to prepare functionalized fabrics, *Nanotechnol. Sci. Appl.*, 10, 147.
77. Vasantharaj, S., Sathiyavimal, S., Saravanan, M., Senthilkumar, P., Gnanasekaran, K., Shanmugavel, M., Pugazhendhi, A. (2019). Synthesis of ecofriendly copper oxide nanoparticles for fabrication over textile fabrics: Characterization of antibacterial activity and dye degradation potential, *J. Photochem. Photobiol. B Biol.*, 191, 143–149.
78. El-Naggar, M.E., Shaheen, T.I., Zaghloul, S., El-Rafie, M.H., Hebeish, A. (2016). Antibacterial activities and UV protection of the in situ synthesized titanium oxide nanoparticles on cotton fabrics. *Ind. Eng. Chem. Res.*, 55, 2661–2668.

79. Jia, L., Huang, X., Tao, Q. (2019). Enhanced hydrophilic and antibacterial efficiencies by the synergetic effect TiO_2 nanofiber and graphene oxide in cellulose acetate nanofibers, *Int. J. Biol. Macromol.*, 132, 1039–1043.
80. Tripathi, R., Narayan, A., Bramhecha, I., Sheikh, J. (2019). Development of multifunctional linen fabric using chitosan film as a template for immobilization of in-situ generated CeO_2 nanoparticles, *Int. J. Biol. Macromol.*, 121, 1154–1159.
81. Yusuf, M. (Ed.) (2018). *Handbook of Renewable Materials for Coloration and Finishing* (John Wiley & Sons, USA).
82. Yusuf, M., Shahid, M. (eds.) (2022). *Emerging Technologies for Textile Coloration*, CRC Press, Boca Raton.
83. Yusuf, M. and Madhu, A. (2022). Smart nanotextiles for filtration. In: Yilmaz, N.D. (ed.), *Smart Nanotextiles: Wearable and Technical Applications* (Scrivener Publishing, USA), 341–360.
84. Hezavehi, E., Shahidi, S., Zolgharnein, P. E. (2015). Effect of dyeing on wrinkle properties of cotton cross-linked by butane tetracarboxylic acid (BTCA) in presence of titanium dioxide (TiO_2) nanoparticles, *Autex Res. J.*, 15, 104–111.
85. Ugur, S.S., Bilgiç, M. (2017). A novel approach for improving wrinkle resistance and flame retardancy properties of linen fabrics, *Bilge Int. J. Sci. Technol. Res.*, 1, 79–86.

Chapter 6

Nanomaterials in Textiles: Performance, Health, and Environmental Aspects

Liliana Indrie,[a] Sabina Gherghel,[a] and Steven McNeil[b]
[a]*University of Oradea, Faculty of Energy Engineering and Industrial Management, Department of Textiles, Leather and Industrial Management, Universității str., no. 1, 410087, Oradea, Romania*
[b]*AgResearch, Textile Group, Lincoln Research Centre, Private Bag 4749, Christchurch, New Zealand*
lindrie@uoradea.ro

Nanotechnology has undergone an impressive evolution over recent decades. In the future it is expected to have an increased impact on both industry and society by penetrating most technological sectors, leading to innovations that will help solve many of the problems that society is facing nowadays. The use of nanotechnology has developed rapidly in the field of textiles, due to its unique and valuable properties it can bring to textiles. Many textile producers employ nanomaterials to give their products new and improved properties, such as greater durability comfort, and hygienic properties, and also to a reduction in the cost of production. Thus, textiles, through the use of nanotechnology, have become multifunctional, and the fabrics

Handbook of Nanofibers and Nanocomposites: Characteristics, Synthesis, and Applications in Textiles
Edited by Mohd Yusuf and Aminoddin Haji

ISBN 978-981-4968-77-5 (Hardcover), 978-1-003-43274-6 (eBook)
www.jennystanford.com

have acquired special functions, including antibacterial protection, UV, as well as water resistance, and ease of cleaning. This review looks at various applications of nanomaterials in the textile industry, how nanomaterials are incorporated into textiles, as well as potential environmental and health aspects.

6.1 Introduction

Nanotechnology is a collective term for nanoscale technological developments and refers to the manipulation or self-assembly of individual atoms, molecules, or atoms in molecules, into structures that create materials and devices with new or different properties than the starting materials [1, 2]. Nanomaterials have unique chemical, physical and mechanical properties and therefore can be used for a wide range of applications such as medicine, control, and analysis equipment as well as in the food industry, the textile industry, the environment, the IT industry, and security purposes [3–4].

Technologies using small or "nanosized" objects existed hundreds of years before the term nanotechnology was coined. For instance, a 2000-year-old hair dyeing recipe shows that the ancient Greeks and Romans used sulfur nanocrystals for turning gray hair black. Nanodimensional particles have also been recently discovered in stained glass windows dating back to the 16th and 17th centuries demonstrating that gold and silver nanoparticles of various sizes were used to selectively reflect red and yellow light.

The first scientific discussion of nanotechnology took place in 1959 when Richard Feynman [5] in his famous speech "There is Plenty of Room at the Bottom" made the first allusion to nanotechnology by describing it as a discipline designed to manipulate smaller and smaller units of matter, allowing the arrangement of atoms to the liking of researchers. Other key events include Norio Taniguchi's first use of the term "nanotechnology" in 1974 [6], and K. Eric Drexler using the term "nanotechnology" in the book *Engines of Creation: The Coming Era of Nanotechnology* in 1986 which is still used in many introductory courses on nanotechnology [7].

In 1985, Harry Kroto, Richard Smalley, and Robert Curlau developed the graphite laser, which led to the discovery of fullerenes [8]. The method of fullerene synthesis was based on the action of the laser on the graphite. The import of this discovery was recognized in 1996 by the award of a Nobel Prize to the three scientists involved.

Although the first fullerene molecules were discovered in 1985, carbon nanotubes only became widely known in 1991. Carbon nanotubes are members of the fullerene family and were discovered in the soot deposited on a carbon cathode of a voltaic arc by Sumio Iijima in NEC research laboratories in Tsukuba, Japan [9].

At present, nanotechnology is continuously evolving, being increasingly diversified, due to the sustained efforts of many teams of researchers around the world. The small size of nanoparticles (below 100 nm), makes them behave significantly differently from larger pieces of materials due to surface effects and quantum effects. The mechanical, optical, electrical, and magnetic properties of the materials are influenced by these factors, as well as their chemical reactivity. Regarding the economic and environmental potential of nanotechnology, explosive economic growth is expected over the next 10 years.

6.2 Nanotechnology in Textiles

In the last decades, we have witnessed a rapid development of nanotechnology in all industries, most notably in the textile industry in the areas of fiber and yarn manufacturing and fabric finishes. The use of nanotechnology in the textile industry [10–13] was reflected by the many publications that describe the application of nanoparticles to textile materials in order to impart distinctive functionalities.

The implementation of nanotechnology in the textile industry has led to increased durability of fabrics over time, enhanced comfort and hygienic properties, and cost reductions. Nanotechnology is used in a wide range of applications in the textile industry: textile finishing, sportswear, fashion, military/security, medical, textile-based sensors, and conductive textiles. This plays an important part in the functionality and performance of nanotextiles and permits textiles to become multifunctional [14, 15].

6.3 Nanoparticles

Given the size of nanomaterials, there are currently three directions for their development: 1D nanomaterials (ultrathin layers), 2D nanomaterials (nanowires, nanotubes), and 3D nanomaterials (nanoparticles). Nanomaterials such as carbon nanotubes, graphene, and various nanoparticles (clay, carbon black, metal, and metal oxide) are used to improve the properties of textiles.

Nanoparticles can be obtained by using synthetic methods that include chemical methods such as chemical precipitation, microemulsion, electrochemical synthesis, thermal decomposition, or physical methods based on microwaves, laser ablation, sonochemical reduction, etc. The method used for synthesis largely determines the characteristics of the nanoparticles (e.g., size, shape, particle size distribution, surface chemistry), which determine their properties which in turn determine the applications to which they are applied. By coating the surfaces with nanoparticles, it is possible to obtain fabrics and clothing with special functions such as UV protection, antibacterial properties, odor reduction, self-cleaning, flame and wrinkle resistance, soil release properties, and water and stain repellent [13].

Nanoparticles fundamentally influence the properties of the materials [16]. By incorporating them into textiles, some properties such as conductivity, strength, shrinkage, and flammability may be affected but their ability to breathe will not be reduced, nor the aesthetic feel [14, 16]. By using conventional finishing approaches, fabrics may lose some functions after washing or wearing. Nanoparticles can provide high durability to treated fabrics because they have a large surface-area-to-volume ratio and large surface energy, showing a better affinity for fabrics.

An important role in determining the adhesion of particles to fibers is their size. Small particles will penetrate deep into the fabrics and adhere strongly to the fabric matrix while large particle agglomerations can be easily removed from the fiber surface [17, 18].

6.3.1 Routes to Incorporating Nanoparticles into Textiles

6.3.1.1 Application of prepared nanoparticles

An important method for preparing nanoparticles is pyrolysis, and it is widely used to make so-called fumed silica and titanium dioxide. Such nanoparticles can be applied to textiles by coating or spraying. One of the most widely studied nanoparticles on textiles is the photoactive fuming titanium dioxide Aeroxide® P25, made by the Evonic company of Germany [19–21]. Nanoparticles can be applied to textile fibers by using different coating techniques (sol-gel, plasma polymerization, layer-by-layer) in order to make the fabric more resistant to extreme climatic conditions and to increase its durability [22].

6.3.1.2 *In situ* preparation

The sol-gel method is widely used to make both nanoparticles and oxide films [23, 24]. The principle of this method is the transformation of the precursors via a sol (a colloidal suspension) and the gradual evaporation of the solvent, followed by heat treatment of drying and crystallization of the gel. The basis of the sol-gel process is hydrolysis reactions and condensation of metal alkoxide precursors or metal salts. Two routes for the sol-gel production of oxide films are used: the alkoxide route in which organo-metallic precursors are used, which are generally more expensive, and the non-alkoxide route (salts) in which aqueous or alcoholic solutions of metal salts are used (acetates, chlorides, nitrates, sulfates) [24].

Obtaining thin films is the most important application of sol-gel processes because the small thickness of the films allows them to easily overcome the volume contractions that occur when the gel dries. The deposition of sol-gel films can be done by the dip-coating technique as well as by the spin-coating technique [24].

The sol-gel method has been well-studied for textile applications due to its advantages compared to the conventional textile finishing process [25]. This technology can be applied to textiles to improve different functional finishes such as flame-retardant [26], anti-mosquito [27] water and oil repellent [28, 29], antibacterial [30, 31] and antiwrinkle processes [32], UV protection [33], self-cleaning [34], and soil-repellency properties [35]. There are the following reasons for using sol-gel technology in textile: is an eco-friendly process, consumes less energy, provide the option to modify the coating's thickness and long-lasting properties of finished fabrics, utilize fewer chemicals, low-temperature treatment, and low toxicity to human health [36].

Nanoparticles can be formed on textiles by methods other than sol-gel. For instance, by reduction of starting materials, after soaking them in the textile, e.g., copper and copper oxide nanoparticles have been made in textiles by reducing copper acetate [37]. In some cases, the reducing agent can be the textile itself, as shown in the preparation of silver nanoparticles on wool [38]. The reduction can also be achieved with ultrasound [39].

6.3.2 The Effect of Nanoparticles on the Handle of Textiles

When nanoparticles are formed on textiles or applied to them ready-formed, there is a possibility of bridges forming between fibers and/or yarns, thus causing an increase in the stiffness of the textile, which is typically observed as an increase in the fabric bending length. For example, *in situ* formation of copper/copper oxide nanoparticles increased the bending lengths of polyester and wool fabrics by 25% and 11%, respectively [37]. Increased fabric stiffness is undesirable for fabrics intended for use in apparel as it reduces the comfort, but it can be desirable for technical applications, such as sun awnings and geotextiles, as is the associated increased tensile strength. Regardless of any bridging, nanoparticles that increase fiber-fiber friction, will also increase friction between the textile and the hands and skins of wearers, thereby reducing softness, a property that is highly valued by consumers [40, 41]. The adverse effect of nanoparticles can be overcome by the co-application of chemical softeners [42–44].

Nanoparticles can offer a better way of functionalizing textiles than the more commonly used continuous coatings. While both approaches can form inter-fiber and inter-yarn bridges, there is an important difference at the fiber level. It has been shown that the discontinuous nature of nanoparticle applications means they have minimal effect on the extension/compression of fiber surfaces that accompany fiber bending. In contrast, continuous coatings must extend/compress in accord with the underlying fiber, which is particularly challenging for fibers such as wool, where extension/compression of the fiber surface is concentrated at the edges of the surface sales, and coatings can cause scale edge "pinning" [45].

6.4 Applications of Nanoparticles in Textiles

With the help of nanotechnology, the properties of fabric can be improved by implementing appropriate techniques for finishing, coating, and/or modifying the surface. Below, we present some applications of nanoparticles to enhance the performance of textiles.

6.4.1 Nanoparticle-Textile Catalysts

Textiles, particularly fabrics, are emerging as a promising support for nanoparticle catalysts, as they prevent them from aggregating while enabling rapid liquid mixing and transfer of reactants and products [46]. Textiles can be employed in a filter-type arrangement where the reactants (liquids or gases) simply pass through a single layer of nanoparticle-containing fabric, or in sophisticated spinning mesh disk reactors, where reactants are pumped onto the center of a rotating fabric [47]. Some examples of published studies in this area are shown in Table 6.1.

Table 6.1 Studies of textile-supported nanoparticle catalysts

Nanoparticle	Reaction catalysed	Reference
Cobalt, manganese, zinc & chromium ferrites	Ugi preparation of benzimidazole, quinazolinone and quinoxaline derivatives	Shaabani, A. & Hezarkhani, Z. (2017) [48]
Gold chloride	Addition of methanol to 3-hexyne	Borrmann, T., et al. (2013) [49]
Iron oxide	synthesis of polyfunctional heterocyclic compounds	Shaabani, A., et al. (2016) [50]
Iron oxide	Peroxidase-like activity	Safarik, I., et al. (2021) [51]
Iron oxide	Heterogeneous Fenton treatment of effluent	Atanasova, D., et al. (2021) [52]
Manganese dioxide	Aerobic oxidation of alkyl arenes, alcohols and sulphides to carbonyl and sulphoxides	Shaabani, A., et al. (2015) [53]
Manganese dioxide	Ugi reaction of aromatic hydrocarbons	Shaabani, A., et al. (2015) [54]
Palladium	Reduction of nitic oxide	Gioria, E., et al. (2021) [55]

6.4.2 Protecting Textiles from Insect Damage

The biodegradability of natural fibers means they a less of a threat to human health and the environment, but it does mean they can be damaged by fungi and insects [57, 58]. Nanotechnology is emerging as a new approach for protecting textiles without recourse to conventional insecticides. A good example of this approach is nanocides; these cause desiccation by rupturing the water-impervious cuticle of insects. This approach is well established for the protection of stored products from insects, and the control of bedbugs and ants in buildings [59], but has only recently received attention with textiles. Silver, zinc oxide, kaolinite, and titanium dioxide nanoparticles have been used as control agents against beetle and moth species such as *Anthrenus verbasci*, *Anthrenocerus australis*, and *Tineola bisselliella* [60–63]. For instance, titanium dioxide had a significant antifeedant effect against the common clothes moth (*Tineola bisselliella*) at levels as low as 0.1% on a mass of wool fabric. When applied to the carpet, titanium dioxide showed statistically significant control of larvae of both *Anthrenocerus australis* and *Tineola bisselliella*, although at levels below that desired by industry and consumers [62]. Further research on this topic may well result in highly effective treatments, as scientists learn more about the most effective nanoparticles to control different species of pests.

6.4.3 Resistance to Dry Soiling

An interesting application of nanoparticles in the textile space makes use of their being of similar size to soil particles. The soils commonly encountered in the street and at home have many components, the most abundant being humus, sand, clay, cement, and silica [64]. Nanoparticles on the surface of fibers, occupy interstices, and cracks (especially on abraded fibers) which would otherwise be sites for the adherence of soil particles. Thus, the nanoparticles make it more difficult for soil particles to remain in the fibers, which reduces the rate of soiling and makes soil removal more efficient, by the washing of fabrics and carpets, and the vacuum cleaning of carpets, see Fig. 6.1.

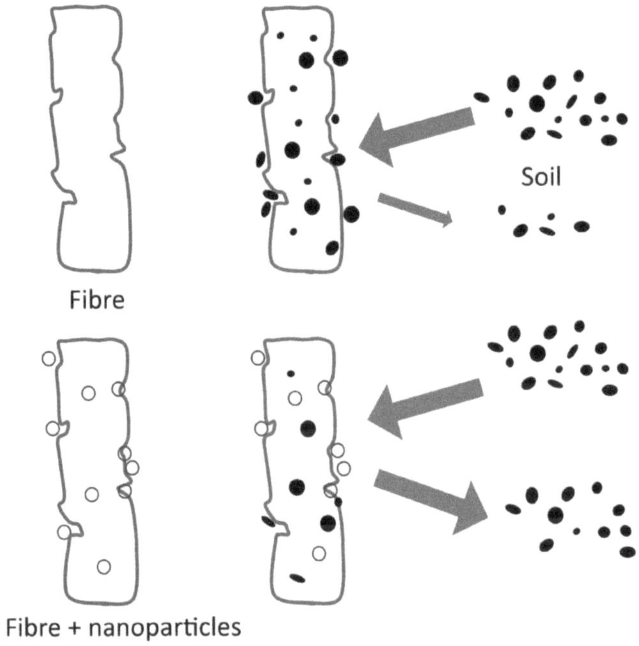

Figure 6.1 Illustration of how nanoparticles can reduce the adherence of soil to fibers.

The dry soiling of automotive textiles (headlining and side linings) by street dust was reduced by applications of hydrophobic silica particles (HDK-C10, Wacker Chemie, München, Germany) with silane binder. The reductions in soiling were maintained over several soiling/washing cycles [65].

The Lanasan NCF (Nano Carpet Finish) process, utilizes silica nanoparticles to block irregularities on the surface of wool fibers [66]. This process can be performed during dyeing or yarn scouring and can reduce the rate of soiling (after vacuum cleaning) by 35% (see Fig. 6.2). Along with other benefits related to increasing fiber-fiber friction (see Section 6.4.7).

The benefits of reduced soiling are considerable, as over the life cycle of a garment, washing uses more energy than production or transportation [67]. Reduced soiling of carpets will prolong their useful lifetime, and reduce the amount of water, energy, and human effort needed to maintain them [68].

Figure 6.2 Carpets made from untreated yarn (left) and Lanasan NCF treated yarn (right) after being subjected to a rapid soiling test (Wools of New Zealand Test Method 267) [66].

6.4.4 Self-Cleaning Applications

In recent years, the attention of many researchers has been focused on the so-called lotus effect. It refers to the self-cleaning property as a result of the superhydrophobicity found in the leaves of the lotus plant. On the surface of the lotus leaf, there is a layer of natural wax, of the order of a few microns thick that imparts a low surface tension on the surface of the leaves and flower petals. Thus, due to the extremely small contact angle, drops of liquid on the surface take on an almost spherical shape, which allows the, to readily slide-off. Such movement of liquids will remove dust or dirt particles from the plant.

Starting from the "lotus effect," the new generation nano-treatments have been perfected in order to impart super-hydrophobicity and self-cleaning properties to surfaces as well as to maintain the nanoparticles on the treated surface. The nano-treatment is applied and then fixed by chemical bonds to the surface of the material that is intended to be protected. Protective nano-treatments preserve the specific properties of the treated materials, thus prolonging their lifespan.

In recent years, many efforts have been made to apply titanium dioxide nanoparticles on textiles in order to produce articles with multifunctional properties such as high photocatalytic activity, non-toxicity, and strong self-cleaning ability. The preparation of TiO_2

nanoparticles is relatively straightforward, however, the low binding efficiency of the TiO_2 nanoparticles to textile creates a problem for the stability and durability of nanocomposite systems during use.

The lotus leaf effect on fabrics was considered in the excellent review of nanostructured titania and titanate materials for environmental and energy applications [69]. Huang et al. proposed a method for multifunctional fabrics with effective self-cleaning, oil-water separation, and anti-UV properties [70]. They developed a TiO_2 fabric composite based on *the self-cleaning* mechanism of *lotus* leaves to prepare marigold flower-like hierarchical TiO_2 particles by a one-pot hydrothermal reaction on cotton fabric. Tudu et al. analyzed the wettability, morphology, and droplet behavior of superhydrophobic cotton fabric and the effect on mechanical, chemical, and thermal properties including resistance to rust, stains, soils, bacteria, and water as well as utility for oil-water separation [71].

6.4.5 UV Protection

Ultraviolet radiation has a detrimental effect not only on the skin but also on clothes. Exposure to ultraviolet radiation can cause sunburn and skin cancer. According to Dhineshbabu et al. [72], protection against ultraviolet radiation has been shown to be necessary for clothing and other textiles.

Lately, many studies have been made on the application of UV-blocking treatments on fabrics by means of nanotechnology. Textiles are predestined as protection against UV radiation, because when suitable materials and constructions are employed, they offer particularly effective and convenient protection against intense UV radiation, in some cases the protection being higher than that of the strongest sunscreens (sun blocker).

Recent studies [73–77] have shown that nanosized TiO_2 and ZnO particles have more efficient performance relative to the organic ultraviolet absorbers to block ultraviolet radiation because they have a much larger surface-area-to-volume ratio. For optimum UV protection, only the outside face of apparel should be covered with nanoparticles, because only this is subject to the action of

UV. The sol-gel method is one of the methods that can be used to apply nanoparticles to the surface of materials. In one of the earlier studies in this area, Xin et al. used the sol-gel method (titanium tetraisopropoxide, hydrolyzed at pH 1–2), to impart UV-blocking treatments to a cotton fabric which was fast to washing (up to the equivalent of 55 home launderings) [78].

6.4.6 Antimicrobial Applications

Contact with bacteria and fungi negatively affects cellular metabolism and inhibits both cell growth and multiplication. The growth of bacteria and fungi can cause infections, odors, itching, and sores. In order to impart antibacterial properties to textiles, nanoparticles of silver, titanium dioxide, and zinc oxide [79–82] have been investigated along with conventional agents such as triclosan and chitosan [83–85].

Silver nanoparticles have captured a great deal of interest and are widely used in consumer products because of their properties. Textiles are treated with silver nanoparticles in order to use the antibacterial and fungicidal properties of silver, particularly to destroy the bacteria responsible for the production of body odors. For instance, Repon et al. reported high activity of silver nanoparticle-treated cotton fabrics against both the gram-positive and gram-negative bacteria [86]. In a similar study, Montazer et al. applied a dispersion of nanosilver to a nylon carpet and observed a 99% reduction in *Staphylococcus aureus* and a 79% reduction in *Escherichia coli*, without affecting the color of the carpet [87].

6.4.7 Other Textile Properties

It has been known for many years that the spinning of yarns can be improved by friction-enhancing materials such as colloidal silica [88]. In the same way, the Lanasan NCF discussed in Section 6.4.3, increased the friction between fibers, thereby reducing the shedding of loose fibers by up to 85% in some yarn/carpet constructions. By increasing the strength of the yarn, Lanasan NCF reduced the frequency of yarn breaks during tufting and weaving and the unwanted machine stoppages that they cause [66].

6.5 Potential Effects of Nanomaterials on Health and the Environment

In addition to the benefits of nanotechnology to textiles and other areas such as improved manufacturing methods, water purification systems, energy systems, materials, etc., the potential risks to the environment, health, and safety must be carefully considered. Otherwise, the side effects of nanoparticles will not be adequately monitored before they are released into the environment and humans are exposed to them. The correct assessment of the risks of nanoparticles to health and the environment involves the analysis of the entire life cycle of these particles, starting from their manufacture, and including storage, distribution, implementation, and disposal of waste resulting from their production and use of end-products. Such an assessment will give the necessary understanding of short- and long-term effects.

Nanoparticles are incorporated into various categories of consumer products, including textiles, cosmetics, electronics, and medicines [89, 90]. The impact of nanotechnology on health is potentially both a positive one, considering the potential of nanotechnological innovations in medicine for the cure of certain diseases, and also a negative one, determined by the possible health hazards generated by exposure to nanomaterials. Therefore, it is important for nanotoxicology research to discover and understand how various factors determine the toxicity of nanomaterials, so that the unwanted effects that might result could be avoided.

Nanoparticles can be absorbed by the human body from the air, water, food, clothing, and medications. Most of these nanoparticles are airborne, can enter the bloodstream, and can be transferred to other organs such as the kidneys, brain, and liver. According to Buzea, C. [91], the diseases associated with nanoparticles are asthma, bronchitis, lung and colon cancer, Parkinson's disease, Alzheimer's disease, and Crohn's disease. Due to the very small size of the materials, they can enter the respiratory tract much more easily than larger particles. The way these nanoparticles behave inside the body is one of the significant problems that will have to be solved in the future [92]. Currently, research is being done on how nanoparticles interact with body systems and how people who come in contact with them may be exposed to nanoparticles in their treatment or industrial use.

Exposure to nanoparticles in everyday life, especially at work, can be analyzed in terms of risk factors:

- in the absence of adequate protection, skin exposure risk may rise while working with nanomaterials in liquid media; during pouring or mixing operations there is a possibility of the formation of drops that can be inhaled;
- in the case of non-closed systems, by generating nanoparticles in the gas phase the chances of aerosol release increase;
- the production or manufacture of nanomaterials, the maintenance of equipment or the handling of finished products, and the cleaning of dust collection systems used to capture nanoparticles involves certain risk factors for the skin and the respiratory system.

To the present date, nanoparticle toxicity data, collected using traditional models and methods, are contradictory and inconsistent. Nanoparticles are nondegradable or slowly degradable, they accumulate in organs, interacting with biological processes in the body, being difficult to generalize the health risks associated with exposure to nanomaterials. It is therefore very important that nanomaterials are subjected to an appropriate toxicity assessment before being used industrially or commercially, each new nanomaterial must be assessed separately and each of its properties must be considered.

Regarding the environmental impact, there are important concerns about the danger represented by nanomaterials and research is being done to see if nanomaterials can cause harmful effects when they reach the landfill and if residues could interfere with animals and plants, causing negative effects. Numerous international research programs assess the environmental impact of waste generated by nanodevices or arising from the production of nanomaterials. These types of waste might be extremely harmful due to their size, they can float in the air and thus can easily penetrate plant and animal cells causing negative effects. For example, nanosized particles from commercial textiles, which are released during their washing, drying, or ironing, have the potential to contaminate water and soil. The production of biocidal nanotextiles can have a negative impact on the aquatic ecosystem when dumped into waters. During the wastewater treatment, some nanotextiles were found to contribute to environmental degradation.

6.6 Conclusions

Nanotechnology has become increasingly present in almost all fields of science due to its potential in improving the effectiveness of some industrial and consumer products. In the future, this could have a major impact on the development of new applications, from diagnosis and treatment to the protection and remediation of the environment.

The implementation of nanotechnology in the textile industry has led to increased durability of fabrics over time, comfort, hygienic properties, and cost reduction. Nanotechnology is used in a wide range of applications in the textile industry: medical uses, textile-based sensors, textile finishing, sportswear, fashion, military/security, and conductive textiles. Nanoparticles are a new promising material for the functionalization of fibers and textiles and due to their properties, they are one of the most useful additives to provide additional functionality to fibers and textiles such as: protecting textiles from insect damage, resistance to dry soiling, self-cleaning applications, and protection from UV.

In addition to the benefits of nanotechnology, the risks related to environmental, health, and safety issues must also be considered so that we do not subsequently face negative effects on health and the environment. Nanoparticles are potentially dangerous because they can be absorbed directly by the circulatory system through the skin and lung membranes. Therefore, all their properties must be taken into consideration, and it is very important that they are subjected to an appropriate toxicity assessment before they are used industrially or commercially.

References

1. Joshi, M. and Bhattacharyya, A. (2011). Nanotechnology: A new route to high-performance functional textiles. *Textile Progress*, 43(3), 155–233. https://doi.org/10.1080/00405167.2011.570027

2. Haque, M. (2019). Nano Fabrics in the 21st century: A review. *Asian Journal of Nanoscience and Materials*, 2(2), 131–148. https://doi.org/10.26655/ajnanomat.2019.3.2
3. Patra, J.K. and Gouda, S. (2013). Application of nanotechnology in textile engineering: An overview, *Journal of Engineering and Technology Research*, 5(5), 104–111.
4. Pal, S., Mondal, S., Das, A., Mondal, D., Panda, B., and Maity, J. (2021). Multifunctional textile fabric and their application. *Journal of Engineering and Technology Research*, 9(6).
5. Feynman, R.P. (1959). There's plenty of room at the bottom. *Engineering and Science*, 23, 22–36.
6. Taniguchi, N., Arakawa, C., and Kobayashi, T. On the basic concept of nano-technology. *Proceedings of the International Conference on Production Engineering*, Tokyo, Japan, 26–29 August, 1974.
7. Drexler, E.K. *Engines of Creation: The Coming Era of Nanotechnology*. Anchor Press, Garden City, New York, USA, 1986.
8. Kroto, H.W., Heath, J.R., O'Brien, S.C., Curl, R.F., and Smalley, R.E. (1985). C60: Buckminsterfullerene. *Nature*, 318, 162–163. Doi:10.1038/318162a0
9. Iijima, S. (1991). Helical microtubules of graphitic carbon. *Nature*, 354, 56–58. Doi: 10.1038/354056a0.
10. Mishra, R. and Militky, J. (eds.), Characterization of nanomaterials in textiles. In: *Nanotechnology in Textiles: Theory and Application*, Woodhead Publishing: Cambridge, UK, 2018.
11. Almeida, L. and Ramos, D. (2017). Health and safety concerns of textiles with nanomaterials. *IOP Conference Series: Materials Science and Engineering*, 254, 102002.
12. Yetisen, A.K., Qu, H., Manbachi, A., Butt, H., Dokmeci, M.R., Hinestroza, J.P., Skorobogatiy, M., Khademhosseini, A., and Yun, S.H. (2016). Nanotechnology in textiles. *ACS Nano*, 10(3), 3042–3068. https://doi.org/10.1021/acsnano.5b08176
13. Srinivas, K. (2016). The role of nanotechnology in modern textiles. *Journal of Chemical and Pharmaceutical Research*, 8(6), 173–180.
14. Jeevani, T. (2011). Nanotextiles: A broader perspective. *Journal of Nanomedicine and Nanotechnology*, 2, 7. Doi: 10.4172/2157-7439.100012
15. Sawhney, A.P.S., Condon, B., Singh, K.V., Pang, S.S., Li, G., and Hui, D. (2008). Modern applications of nanotechnology in textiles. *Textile Research Journal*, 78, 731–739.

16. Leavline, E.J., Singh, D.A.A.G., Prasannanayagi, S., and Kiruthika, R. (2015). A compendium of nano materials and their applications in smart nano textiles. *Research Journal of Nanoscience and Nanotechnology*, 5(2), 44–59. Doi: 10.3923/rjnn.2015.44.59
17. Wong, Y.W.H., Yuen, C.W.M., Leung, M.Y.S., Ku, S.K.A., and Lam, H.L.I. (2006). Selected applications of nanotechnology in textiles. *AUTEX Research Journal*, 6(1), 1–8.
18. Haque, M. (2019). Nano Fabrics in the 21st century: A review. *Asian Journal of Nanoscience and Materials*, 2(2), 131–148.
19. Kowal, K., et al. (2014). Biocidal effect and durability of nano-TiO_2 coated textiles to combat hospital acquired infections. *RSC Advances*, 4(38), 19945–19952. https://doi.org/10.1039/C4RA02759K
20. Kowalczyk, D., Brzeziński, S., and Kamińska, I. (2018). Multifunctional nanocoating finishing of polyester/cotton woven fabric by the sol-gel method. *Textile Research Journal*, 88(8), 946–956. https://doi.org/10.1177/0040517517693979
21. McNeil, S.J. and Sunderland, M.R. (2016). The nanocidal and antifeedant activities of titanium dioxide desiccant towards wool-digesting *Tineola bisselliella* moth larvae. *Clean Technologies and Environmental Policy*, 18(3), 843–852. https://doi.org/10.1007/s10098-015-1060-4
22. Joshi, M. and Bhattacharyya, A. (2011). Nanotechnology: A new route to high-performance functional textiles. *Textile Progress*, 43(3), 155–233. https://doi.org/10.1080/00405167.2011.570027
23. Danks, A.E., Hal, S.R., and Schnepp, Z. (2016). The evolution of 'sol–gel' chemistry as a technique for materials synthesis. *Materials Horizons*, 3, 91–112.
24. Muşat, V. *Filmesubţirimultifuncţionale.* Editura Cermi, Iaşi, 2007.
25. Ismail, W.M.W. (2016), Sol–gel technology for innovative fabric finishing: A review. *Journal of Sol-Gel Science and Technology*, 78, 698–707. Doi: 10.1007/s10971-016-4027-y
26. Malucelli, G. (2020). Sol-gel and layer-by-layer coatings for flame-retardant cotton fabrics: Recent advances. *Coatings*, 10, 333. https://doi.org/10.3390/coatings10040333
27. Mulyani, W.E. and Sunendar, B. (2013). Synthesis and characterization of silica-lavender microencapsulation by sol-gel emulsion method for anti mosquito textile. *Advanced Materials Research*, 789, 215–218. https://doi.org/10.4028/www.scientific.net/amr.789.215
28. Mahltig, B. and Böttcher, H. (2003). Modified silica sol coatings for water-repellent textiles. *Journal of Sol-Gel Science and Technology*, 27, 43–52.

29. Shibuichi, S., Yamamoto, T., Onda, T., Tsujii, K. (1998). Super water- and oil-repellent surfaces resulting from fractal structure. *Journal of Colloid and Interface Science*, 208, 287–294.

30. Mahltig, B., Fiedler, D. and Böttcher, H. (2004). Antimicrobial sol-gel coatings. *Journal of Sol-Gel Science and Technology*, 32, 219–222.

31. Rivero, P.J. and Goicoechea, J. Sol-gel technology for antimicrobial textiles. In: Sun, G. (ed.), *Antimicrobial Textiles.* Woodhead Publishing, Sawston, UK, 2016, 47–72. ISBN 9780081005859.

32. Huang K.S., Nien Y.H., Hsiao K.C., and Chang Y.S. (2006). Application of DMEU/SiO_2 gel solution in the antiwrinkle finishing of cotton fabrics. *Journal of Applied Polymer Science*, 102(5), 4136–4143. Doi: 10.1002/app.24246

33. Vihodceva, S. and Kukle, S. (2013). Improvement of UV protection properties of the textile from natural fibres by the sol-gel method. *IOP Conference Series: Materials Science and Engineering*, 49, 12022.

34. Gupta, K.K., Jassal, M., and Agrawal, A.K. (2008). Sol-gel derived titanium dioxide finishing of cotton fabric for self cleaning. *Indian Journal of Fibre and Textile Research*, 33, December, 443–450.

35. Mahltig, B., Audenaert, F., and Böttcher, H. (2005). Hydrophobic silica sol coatings on textiles-the influence of solvent and sol concentration. *Journal of Sol-Gel Science and Technology*, 34, 103–109.

36. Periyasamy, A.P., Venkataraman, M., Kremenakova, D., Militky, J., and Zhou, Y. (2020). Progress in sol-gel technology for the coatings of fabrics. *Materials*, 13, 1838. https://doi.org/10.3390/ma13081838

37. Rezaie, A.B., Montazer, M., and Rad, M.M. (2018). Scalable, eco-friendly and simple strategy for nano-functionalization of textiles using immobilized copper-based nanoparticles. *Clean Technologies and Environmental Policy*, 20(9), 2119–2133. https://doi.org/10.1007/s10098-018-1596-1

38. Mowafi, S., Kafafy, H., Arafa, A., Haggag, K., and Rehan, M. (2018). Facile and environmental benign in situ synthesis of silver nanoparticles for multifunctionalization of wool fibers. *Environmental Science and Pollution Research*, 25(29), 29054–29069. https://doi.org/10.1007/s11356-018-2928-8

39. Abramova, A.V., et al. (2021). Strong antibacterial properties of cotton fabrics coated with ceria nanoparticles under high-power ultrasound. *Nanomaterials*, 11(10), 2704. https://doi.org/10.3390/nano11102704

40. Mahar, T.J. and Wang, H. (2010). Measuring fabric handle to define luxury: An overview of handle specification in next-to-skin knitted

fabrics from Merino wool. *Animal Production Science*, 50(11–12), 1082–1088. https://doi.org/10.1071/AN10119

41. Matsudaira, M. and Matsui, M. (1992). Changes in the mechanical properties and fabric handle of polyester-fibre fabrics through the finishing stages. *Journal of the Textile Institute*, 83(1), 144–155. https://doi.org/10.1080/00405009208631184
42. Hao, L., Gao, T., Xu, W., Wang, X., Yang, S., and Liu, X. (2016). Preparation of crosslinked polysiloxane/SiO_2 nanocomposite via in-situ condensation and its surface modification on cotton fabrics. *Applied Surface Science*, 371, 281–288. https://doi.org/10.1016/j.apsusc.2016.02.204
43. Pithanthanakul, U., Vatanyoopaisarn, S., Thumthanaruk, B., Puttanlek, C., Uttapap, D., Kietthanakorn, B., and Rungsardthong, V. (2021). Encapsulation of fragrances in zein nanoparticles and use as fabric softener for textile application. *Flavour and Fragrance Journal*, 36(3), 365–373. https://doi.org/10.1002/ffj.3648
44. Sugimura, M., McNeil, S. and Zaitseva, L. A study of the interactions of softening agents and nanoparticles with respect to the handle of wool fabrics. *45th Textile Research Symposium*, Kyoto Institute of Technology, Kyoto, Japan, 14–16th September 2017. https://doi.org/10.13140/RG.2.2.29138.38084
45. McNeil, S.J. and Standard, O.C. (2017). Increased bending rigidity of wool fabric imparted by hybrid organic-inorganic sol-gel coatings. *Textile Research Journal*, 87(5), 607–616. https://doi.org/10.1177/0040517516635990
46. McNeil, S.J., Sunderland, M.R. and Leighs, S.J. (2017). The utilisation of wool as a catalyst and as a support for catalysts. *Applied Catalysis A: General*, 541, 120–140. http://dx.doi.org/10.1016/j.apcata.2017.04.
47. Shivaprasad, P., Jones, M.D., Frith, P., and Emanuelsson, E.A.C. (2020). Investigating the effect of increasing cloth size and cloth number in a spinning mesh disc reactor. *Chemical Engineering and Processing: Process Intensification*, 147, 107780. https://doi.org/10.1016/j.cep.2019.107780
48. Shaabani, A. and Hezarkhani, Z. (2017). Ferrite nanoparticles supported on natural wool in one-pot tandem oxidative reactions: strategy to synthesize benzimidazole, quinazolinone and quinoxaline derivatives. *Applied Organometallic Chemistry*, 31(1), e3542. https://doi.org/10.1002/aoc.3542
49. Borrmann, T., Lim, T.H., Cope, H., Lucas, K., and Lorden, M. (2013). Gold nanoparticles on wool in a comparative study with molecular gold catalysts. *Gold Bulletin*, 46(1), 13–18. https://doi.org/10.1007/s13404-012-0076-3

50. Shaabani, A., Hezarkhani, Z., and Faroghi, M.T. (2016). Wool-SO_3H and nano-Fe_3O_4@ wool as two green and natural-based renewable catalysts in one-pot isocyanide-based multicomponent reactions. *Monatshefte für Chemie-Chemical Monthly*, 147(11), 1963–1973. https://doi.org/10.1007/s00706-016-1717-7

51. Safarik, I., et al. (2021). Cotton textile/iron oxide nanozyme composites with peroxidase-like activity: Preparation, characterization, and application. *ACS Applied Materials and Interfaces*, 13(20), 23627–23637. https://doi.org/10.1021/acsami.1c02154

52. Atanasova, D., Staneva, D., and Grabchev, I. (2021). Textile with a hydrogel and iron oxide nanoparticles for wastewater treatment after reactive dyeing. *Journal of Applied Polymer Science*, 138(10), 49954. https://doi.org/10.1002/app.49954

53. Shaabani, A., Hezarkhani, Z., and Badali, E. (2015). Wool supported manganese dioxide nano-scale dispersion: A biopolymer-based catalyst for the aerobic oxidation of organic compounds. *RSC Advances*, 5(76), 61759–61767. https://doi.org/10.1039/C5RA10522F

54. Shaabani, A., Hezarkhani, Z., and Badali, E. (2015). One-pot oxidative Ugi-type three-component reaction of aromatic hydrocarbons of petroleum naphtha. *RSC Advances*, 5(112), 91966–91973. https://doi.org/10.1039/C5RA16608J

55. Gioria, E., Signorini, C., Taleb, M. C., Mihályi, M., and Gutierrez, L. (2021). Palladium nanoparticles on modified cellulose as a novel catalyst for low temperature gas reactions. *Cellulose*, 28, 9135–9147. https://doi.org/10.1007/s10570-021-04118-9

56. Majdoub, M., Amedlous, A., Anfar, Z., and Moussaoui, O. (2021). MoS_2 nanosheets/silver nanoparticles anchored onto textile fabric as "dip catalyst" for synergistic p-nitrophenol hydrogenation. *Environmental Science and Pollution Research*, 28(45), 64674–64686. https://doi.org/10.1007/s11356-021-14882-7

57. Marsh, P.B. and Bollenbacher, K. (1949). The fungi concerned in fiber deterioration. *Textile Research Journal*, 19(6), 313–324. https://doi.org/10.1177/004051754901900601

58. Collie, S.R., Ranford, S.L., Fowler, I.J., and Brorens, P.H. (2019). Microfibre Pollution: What's the Story for Wool? *Proceedings of the 19th World Textile Conference: Autex 2019*, paper 4A3_0151.

59. Lilly, D.G., Webb, C.E., and Doggett, S.L. (2016). Evidence of tolerance to silica-based desiccant dusts in a pyrethroid-resistant strain of *Cimex lectularius* (Hemiptera: Cimicidae). *Insects*, 7(4), 74. https://doi.org/10.3390/insects7040074

60. Ki, H.Y., Kim, J.H., Kwon, S.C., and Jeong, S.H. (2007). A study on multifunctional wool textiles treated with nano-sized silver. *Journal of Materials Science*, 42(19), 8020–8024. https://doi.org/10.1007/s10853-007-1572-3

61. Nazari, A., Montazer, M., and Dehghani-Zahedani, M. (2013). Nano TiO_2 as a new tool for mothproofing of wool: Protection of wool against *Anthrenus verbasci*. *Industrial and Engineering Chemistry Research*, 52(3), 1365–1371. https://doi.org/10.1021/ie302187c

62. Sunderland, M.R. and McNeil, S.J. (2017). Protecting wool carpets from beetle and moth larvae with nanocidal titanium dioxide desiccant. *Clean Technologies and Environmental Policy*, 19, 1205–1213. https://doi.org/10.1007/s10098-016-1297-6

63. Jose, S., Nachimuthu, S., Das, S., and Kumar, A. (2018). Moth proofing of wool fabric using nano kaolinite. *Journal of the Textile Institute*, 109(2), 225–231. https://doi.org/10.1080/00405000.2017.1336857

64. Sanders H.L. and Lambert J.M. (1950). An approach to a more realistic cotton detergency test. *Journal of the American Oil Chemists' Society*, 27, 153–159. https://doi.org/10.1007/BF02634409

65. Klinkhammer, K., Rohleder, E., Graßmann, C., and Janssen, E. (2017). Structured textile surfaces for easy-to-clean properties towards dry soil. *Materials Today: Proceedings*, 4, S101–S106.

66. Ingham, P.E., Sunderland, M.R., McNeil, S.J., and Marazzi R. (2006). Lanasan NCF: Nanoparticles enhance carpet performance. *International Dyer*, 191(1), 23–25. https://doi.org/10.13140/2.1.4142.1447

67. Laitala, K., Boks, C., and Klepp, I.G. (2011). Potential for environmental improvements in laundering. *International Journal of Consumer Studies*, 35(2), 254–264. https://doi.org/10.1111/j.1470-6431.2010.00968.x

68. McCall, R.A. and McNeil, S.J. (2007). Comparison of the energy, time and water usage required for maintaining carpets and hard floors. *Indoor and Built Environment*, 16(5), 482–486. https://doi.org/10.1177/1420326X07082781

69. Zhang, Y., et al. (2015). Titanate and titania nanostructured materials for environmental and energy applications: A review. *RSC Advances*, 5(97), 79479–79510. https://doi.org/10.1039/c5ra11298b

70. Huang, J.Y., et al. (2015). Robust superhydrophobic TiO_2@ fabrics for UV shielding, self-cleaning and oil–water separation. *Journal of Materials Chemistry A*, 3(6), 2825–2832. https://doi.org/10.1039/C4TA05332J

71. Tudu, B. K., Sinhamahapatra, A., and Kumar, A. (2020). Surface modification of cotton fabric using TiO_2 nanoparticles for self-cleaning, oil–water separation, antistain, anti-water absorption, and antibacterial properties, *ACS Omega*, 5(14), 7850–7860, Doi: 10.1021/acsomega.9b04067

72. Dhineshbabu, N.R., and Bose, S. (2019). UV resistant and fire retardant properties in fabrics coated with polymer-based nanocomposites derived from sustainable and natural resources for protective clothing application. *Composites Part B: Engineering*, 172, 555–563.

73. Saleem, H. and Zaidi, S.J. (2020). Sustainable use of nanomaterials in textiles and their environmental impact. *Materials,* 13, 5134. https://doi.org/10.3390/ma13225134

74. Sedighi, A., Montazer, M., and Mazinani, S. (2018). Fabrication of electrically conductive superparamagnetic fabric with microwave attenuation, antibacterial properties and UV protection using PEDOT/magnetite nanoparticles. *Materials & Design*, 160, 34–47.

75. Fouda, A., Saad, E.L., Salem, S.S., and Shaheen, T.I. (2018). *In-Vitro* cytotoxicity, antibacterial, and UV protection properties of the biosynthesized zinc oxide nanoparticles for medical textile applications. *Microbial Pathogenesis*, 125, 252–261.

76. Attia, N.F., Moussa, M., Sheta, A.M., Taha, R., and Gamal, H. (2017). Synthesis of effective multifunctional textile based on silica nanoparticles. *Progress in Organic Coatings*, 106, 41–49.

77. Attia, N.F., Moussa, M. Sheta, A.M., Taha, R., and Gamal, H. (2017). Effect of different nanoparticles based coating on the performance of textile properties. *Progress in Organic Coatings*, 104, 72–80. https://doi.org/10.1016/j.porgcoat.2016.12.007

78. Xin, J.H., Daoud, W., and Kong, Y.Y. (2004). A new approach to UV-blocking treatment for cotton fabrics. *Textile Research Journal*, 74(2), 97–100. https://doi.org/10.1177/004051750407400202

79. Pulit-Prociak, J. and Banach, M. (2016). Silver nanoparticles: A material of the future...? *Open Chemistry*, 14(1), 76–91. https://doi.org/10.1515/chem-2016-0005

80. Abu-Qdais, H.A., Abu-Dalo, M.A., Hajeer, Y.Y. (2021). Impacts of nanosilver-based textile products using a life cycle assessment. *Sustainability*, 13, 3436. https://doi.org/10.3390/su13063436

81. Shan, A.Y., Ghazi, T.I.M., and Rashid, S.A. (2010). Immobilisation of titanium dioxide onto supporting materials in heterogeneous photocatalysis: A review. *Applied Catalysis A: General*, 389, 1–8. http://dx.doi.org/10.1016/j.apcata.2010.08.053

82. Tania, I.S. and Ali, M. (2021). Coating of ZnO nanoparticle on cotton fabric to create a functional textile with enhanced mechanical properties. *Polymers*, 13, 2701. https://doi.org/10.3390/polym13162701

83. Orhan, M., Kut, D., and Gunesoglu, C. (2007). Use of triclosan as antibacterial agent in textiles. *Indian Journal of Fibre and Textile Research*, 32(1), 114–118.

84. Indrie, L., Bonet-Aracil, M., Ilieş, D.C., Albu, A.V., Ilieş, G., Herman, G.V., Baias, Ş., and Costea, M. (2021). Heritage ethnographic objects: Antimicrobial effects of chitosan treatment. *Industria Textila*, 72(3), 284–288. http://doi.org/10.35530/IT.072.03.1812

85. Bou-Belda, E., Indrie, L., Ilieş, D.C., Hodor, N., Berdenov, Z., Herman, G., Caciora, T. (2020). Chitosan: A non-invasive approach for the preservation of historical textiles. *Industria Textila*, 2020, 71(6), 576–579. http://doi.org/10.35530/IT.071.06.1756

86. Repon, M.R., Islam, T., Sadia, H.T., Mikučionienė, D., Hossain, S., Kibria, G., and Kaseem, M. (2021). Development of antimicrobial cotton fabric impregnating AgNPs utilizing contemporary practice. *Coatings,* 11, 1413. https://doi.org/10.3390/coatings11111413

87. Montazer M., Hajimirzababa H., Rahimi M.K., and Alibakhshi S. (2012). Durable anti-bacterial nylon carpet using colloidal nano silver. *Fibres and Textiles in Eastern Europe*, 93(4), 96–101.

88. Murray, E.A., Moore, J.W., and Williams, S. (1947). Studies of the treatment of cotton in sliver form I. Effect of Syton: Spinning studies. *Textile Research Journal*, 17(6), 331–339. https://doi.org/10.1177/004051754701700604

89. Medina C., Santos-Martinez M.J., Radomski A., Corrigan O.I., and Radomski M.W. (2007). Nanoparticles: Pharmacological and toxicological significance. *British Journal of Pharmacology*, 150, 552–558. https://doi.org/10.1038/sj.bjp.0707130

90. Maynard, A. and Michelson, E. (2006). *The Nanotechnology Consumer Products Inventory*. Woodrow Wilson International Center for Scholars.

91. Buzea, C., Pacheco, I., and Robbie, K. (2007). Nanomaterials and nanoparticles: Sources and toxicity. *Biointerphases*, 2, MR17–MR72.

92. Morones J.R., Elechiguerra J.L., Camacho A., Holt K., Kouri J.B., Ramirez J.T., and Yacaman M.J. (2005). The bactericidal effect of silver nanoparticles. *Nanotechnology*, 16, 2346–2353. https://doi.org/10.1088/0957-4484/16/10/059.

Chapter 7

Overview of Polymer/Metal-Oxide Nanocomposites: Synthesis, Properties, and Their Potential Applications

Bilal Ahmed,[a] Arvind Singh,[b] Ruhinaz Ushal,[c] Mohd Yusuf,[a] Sachin Kumar,[d] Animesh Kumar Ojha,[e] and Wasim Khan[f]

[a]*Department of Natural and Applied Sciences, School of Science and Technology, The Glocal University, Mirzapur Pole, Saharanpur, 247121, India*

[b]*Department of Physics, Indian Institute of Technology, Patna, 801106, India*

[c]*School of Pharmacy, The Glocal University, Mirzapur Pole, Saharanpur, 247121, India*

[d]*School of Chemical Engineering, Yeungnam University, Gyeongsan, Gyeonbuk, 38541, Republic of Korea (South Korea)*

[e]*Department of Physics, Motilal Nehru National Institute of Technology Allahabad, Prayagraj, 211004, India*

[f]*Department of Petroleum Engineering, School of Science and Technology, The Glocal University, Mirzapur Pole, Saharanpur, 247121, India*

b.ahmedphy@gmail.com

This chapter aims to give a concise overview of recent research considering the synthesis, properties, and applications of polymer/metal-oxide composites of nanostructures. Nowadays polymer/metal-oxide nanocomposite assemblies have been attracting

Handbook of Nanofibers and Nanocomposites: Characteristics, Synthesis, and Applications in Textiles
Edited by Mohd Yusuf and Aminoddin Haji

ISBN 978-981-4968-77-5 (Hardcover), 978-1-003-43274-6 (eBook)
www.jennystanford.com

enormous attention from the scientific community, because of their exceptional physico-chemical characteristics, chemical tenability and which give high hopes for diversified applicability with respect to new trendy science and technology vistas. In addition, polymer-based nanocomposites can be considered attractive aspects. The synchronous applications have provided new opportunities in many fields, including textile, medical, environmental, energy evolution reaction, and energy storage supercapacitors. This chapter encompasses the overview of the synthetic techniques, characteristics/properties, and applications of polymer/metal oxide from a material science perspective.

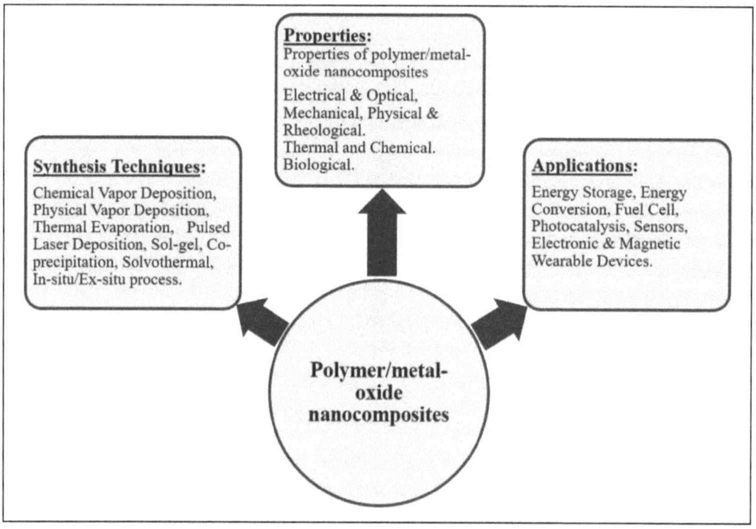

Figure 7.1 Schematic presentation of polymer/metal-oxide nanostructure.

7.1 Introduction

In recent decades, the quick progress of nanoscience and technology has led to the development in the fields of nanomaterials sciences because of its outstanding physical, optical and chemical properties [1, 2]. New materials and the synthesis procedures of nanostructured materials could help in the tremendous surge in the preservation and control the natural resources and energy consumption [3, 4]. Though polymer/metal-oxide nanocomposites have gained high attention due to their irreplaceable mechanical, magnetic, electrical, optical, catalytic, and chemical properties, having a wide range of scientific applications, i.e., fuel cells, super capacitive electrodes, water splitting, sensors, medical, pyroelectric, anticorrosion coatings, flame retarding, photocatalysis, etc. [5–14].

In the past few years, metal oxides such as CdO, ZnO, TiO_2, ZrO_2, CuO, WO_3, MoO_3, SiO_2, NiO, etc. have different morphological structures i.e., spherical, tubes, rods, shells, sheets, etc. have been regarded as potential candidates by the scientific community, owing to their unique physical, electronic, optical and chemical features, owing to these features these materials are used in different field of science and technology for various applications [15–42]. Besides, the given features of the metal-oxide nanostructures, the scientific community is very curious, to change the structural and physical properties of the metal-oxide nanostructures by introducing polymer nanostructures. In order, to synthesize the polymer/metal-oxide nanostructures with separated phase structures. With the steady change in size and structure of metal-oxide nanoparticles, the incorporation of polymer matrix includes a variation in the dimension of the nanocomposites that alters the physical and chemical properties of the nanocomposites. Now, it's a very typical issue to have a homogeneous dispersion of metal oxides throughout the matrix of the polymer to produce a uniform alignment of the nanostructure. Due to the incorporation of metal-oxide nanostructured materials the properties of the polymer can be enhanced in terms of stability, tensile strength, toughness, glass transition temperature, optical, chemical, and many more properties are found to be enhanced by the incorporation of metal-oxide nanostructures with diverse polymer matrixes [43]. Thus, polymer with metal-oxide nanocomposites has been used in different scientific applications such as supercapacitors electrode materials, photocatalysis, water splitting, antibacterial, flame retardancy, medical, textile coatings, and many other applications [44–53]. Nowadays a large number of research articles have been published related to metal-oxide and polymer composites, and it's quite typical to go through all research articles. Therefore, this chapter would provide a general overview and basic synthesis techniques involved in the synthesis of polymer and metal-oxide non-structural composites, including the enhancement in the structural, optical, chemicals, and other related properties with their evaluation of different applications in the field of science and technology.

7.2 Synthetic Techniques

There are different techniques for the synthesis of polymer, metal oxide, and polymer/metal-oxide nanocomposites with efficient surface morphological structure, and novel properties. To overwhelm these issues, scientists have drawn considerable attention to the synthesis process and developed new combined methods with higher control on the morphology and novel properties. Therefore, various physical and chemical synthesis approaches such as PVD, thermal evaporation, pulsed laser deposition (PLD), spray pyrolysis, CVD, sol-gel, hydrothermal, co-precipitation, etc. are employed for the production of polymer/metal-oxide nanocomposites are discussed in the following sections.

7.2.1 Physical Vapor Deposition

Physical vapor deposition (PVD) [53] is a synthesis technique in which a sample is placed in a closed chamber and evaporates above the melting point, where the evaporated particles follow a free route formed by a vacuum and deposit on the substrate [54].

7.2.2 Thermal Evaporation

Thermal evaporation is a basic technique for the preparation of nanostructures [55]. In the thermal evaporation techniques, different components are required such as a source container, furnace, vacuum unit, and substrates for precise uniformity control [56]. The high-temperature furnace is used to vaporize the sample materials, for the deposition of thin film nanostructures. Uniformity, grain structure, thickness, tension, adhesion strength, electrical, and optical properties can all be controlled using this technology [57].

7.2.3 Pulsed Laser Deposition

In this technique, the required material is vaporized using a laser with narrow frequency bandwidth and a high-power density

[58]. Various nanostructures (such as nanopowders, nanotubes, nanorods, quantum dots, etc.) have been synthesized via PLD techniques [13]. However, the previous techniques failed to make or were troublesome for the deposition. The versatility of PLD comes from the fact that there is no constraint on the target material. Matrix-assisted PLD is employed for hybrid metal organics, polymer composites, etc. The thickness of the depositing materials ranges from 10–1000 nm on the various substrates [59]. It is reported that PLD was used to synthesize TiO_2 nanoparticles-based film that may be used for a gas sensor, energy storage, and other applications, where the synthesized nanoparticles were measured 10–100 nm in diameter [60].

7.2.4 Chemical Vapor Deposition (CVD)

In the chemical vapor deposition (CVD) techniques, precursor gases pass toward one or more heated substrates, and chemical reactions take place on the substrate's surface, resulting in the deposition of thin films over the substrate [61]. Thereafter, an exhaust chamber is used to collect the by-products generated during the chemical reaction process. In this technique, high-speed electrons, ions, plasmas, or hot filaments are used to carry out the entire process at temperatures ranging from 200–1200°C [61, 62]. The physical and chemical properties of a deposited film depend on the growing morphology, crystallinity, and nucleation growth. Over the last few years, the use of CVD to make high-purity metal-oxide nanostructures has been reported 18. The use of CVD to make high-purity zinc oxide nanoparticles has also been reported. The scientists discovered that the heterocubane clusters as a precursor for the synthesis of ZnO nanoparticles, with breakdown achieved by removing methane and propane [63]. Because of the creation of the Zn_4O_4 cluster, the primary particles combine and form bigger particles. In 2012, Amara et al. [64] used the standard CVD approach to synthesize magnetite nanoparticles. The scientists used ferrocene as a precursor for making magnetite nanospheres and nanocubes on a polyvinyl pyrrolidone substrate.

7.2.5 Sol-Gel Technique

This approach is generally used for the fabrication of nanostructured materials by using colloidal dispersion or alkoxides with the intermediate route sol and gel at elevated temperatures [65–67]. In 1990, Hench et al. [68] reported that the sol-gel process includes homogeneous multicomponent oxides at an atomic level. In this process, nucleation and particle growth are likely to be controlled to provide the proper size and shape. This process can be used for the synthesis of all types of oxide composition, hybrid materials composites of organic and inorganic materials. Besides, the limitation is not an economical process, because in this process very expensive precursors and alkoxides are required.

7.2.6 Co-precipitation Techniques

A co-precipitation is a chemical approach for the synthesis of metal-oxide nanostructures and their composites with different compounds [69]. In this approach, the foremost parameters (i.e., solution concentration, temperature, pH, viscosity, surface tension, duration, and speed of stirring) control the size, surface morphology, and other properties of the product [70]. During the synthesis, two types of micelles (normal and reverse) get generated. A large amount of wastage affects the production yields and is the main cause to limit the use of this technique [71–73].

7.2.7 Solvothermal Techniques

In the solvothermal process, the precursor material is placed in a closed teflon-lined stainless-steel chamber under the condition of high-pressure and temperature technique, and a precursor undergoes the chemical reaction in the presence of a solvent. Solvent plays a decisive role in the synthesis of nanostructures. We report the synthesis of WO_3 nanostructures with different morphology by using ammonium nitrate as a base and deionized water as a solvent [15, 18]. Similarly, in 2011 Ye et al. [74] synthesize monocrystal TiO_2 nanostructures using tetra-butyl titanate and acetic acid). In the solvothermal process, we can easily control the physicochemical properties of the synthesized products by varying parameters (i.e., thermodynamically and chemically).

7.2.8 *Ex/In situ* Formation

The synthesis of metal-oxide nanostructures into the polymer matrixes can be done using two different processes, i.e., *ex situ* and *in situ* processes [43, 75].

7.2.8.1 *Ex situ* process

In the *ex situ* process, the inorganic nanoparticles are first synthesized, then incorporate into the polymer matrix, based on the physical incorporation of metal-oxide nanostructures into the polymer matrixes [43]. However, in this approach, the non-homogeneous dispersion of metal-oxide nanostructures is a very typical issue. To overcome this issue, the *in situ* approach can be used. In this approach/process, metal-oxide nanostructures are generated inside the polymer matrixes using the precursor, which is converted to the desirable nanostructures.

7.2.8.2 *In situ* process

The *in situ* process [75] is one of the most successful chemical processes and allows the incorporation of metal-oxide nanostructures in the polymer matrix with control over the particle size and desired morphology. During the synthesis process polymer act as a reaction mediator where the metal-oxide nanostructures are generated via the chemical conversion of the metal precursor.

7.2.9 Processing of Polymer/Metal-Oxide Nanocomposites

The researcher's curiosity has been piqued by the mutual characteristics of polymer/metal oxide. The hybrid composites exhibit improved conductivity and thermal characteristics as well as high elastic stiffness and wear resistance. The crucial variables for producing the useful properties of the nanocomposites are the volume fraction of the constituents, together with the interfacial exchanges between the matrixes and nanoparticles. Effective dispersion of nanoparticles in the polymer matrix is always difficult owing to their tendency to aggregate [76, 77].

There are three typical approaches for creating polymer/ metal-oxide nanocomposites. The **first** technique involves *in situ* polymerization of monomers with metal-oxide nanoparticles. The **second** method involves mixing melt or solution-based metal-oxide nanoparticles and polymer directly. The sol-gel procedure is the **third**. In 2000, hydrophobic TiO_2 nanoparticles with polybutylene succinimide diethyl triamine modifications were created, and 5 wt% of these particles were disseminated in styrene before the mini-emulsion procedure (a form of *in situ* polymerization technique) [78]. Polystyrene was used to encapsulate nanoparticles to a degree of about 89 % weight [79].

The condensation and hydrolysis reactions make up the process' two primary reactions. Condensation is based on the creation of –O–M–O– bonds, whereas hydrolysis requires the breaking of the organic chain bonded to the metal and replacement with –OH groups through nucleophilic addition [68]. The key variables are the solvent, water content, temperature, and metal reactivity. Materials created using sol-gel processing exhibit great purity, homogeneity, and low sintering temperatures as compared to reactions that use more traditional methods [80].

The ***ex situ*** method of blending, which can be separated into solution blending and melt blending, is the most straightforward way to create metal-oxide/polymer nanocomposites. The poly (butylene succinate)/TiO_2 nanocomposites were made using melt blending with an extruder as the processing tool [81]. In relation to the dispersion state of TiO_2 nanoparticles, the scientists investigated their photo-induced breakdown and biodegradability. While melt processing is still preferred by industry because of its low cost and ease of usage in large-scale manufacturing for commercial applications, solution blending is employed more frequently in laboratories than melt blending for the processing of nanocomposites.

7.3 Properties of Polymer/Metal Oxide–Based Nanocomposites

Owing to their special qualities, metal-oxide nanoparticles are crucial in a variety of applications, like the treatment of wastewater, tissue engineering, MRI, paints, and cancer treatment.

7.3.1 Physical, Mechanical, and Rheological Properties

The incorporation of nanoparticles into the polymer matrix improves the physical, mechanical, and rheological properties. The coefficient of thermal expansion, Young's and shear modulus, electrical and thermal conductivity, and chemical activity are just a few of the properties that change depending on the size and form of the nanoparticles. The produced nanocomposites' reactivity, as well as their electrical and optical properties, are impacted by the increased surface energy of nanoparticles, which alters the crystallographic structure [82]. Rheological characterization of polymeric nanocomposites is crucial because it provides a general understanding of addition of different fillers affects the structure connection, which is crucial for comprehending the viscoelastic flow behavior of the system [83]. The structural and dielectric characteristics of isotactic polypropylene- and iron oxide nanoparticle-based polymer nanocomposites have been studied by [84].

Liang et al. [85] explored the fluorescent epoxy nanocomposites reinforced with PDA-functionalized zinc oxide (ZnO) nanoparticles. The results show that the PDA-functionalized ZnO nanoparticles were diffused uniformly in the epoxy matrix. Tensile strength measurements revealed that the epoxy nanocomposites containing PDA-functionalized ZnO nanoparticles had a higher tensile strength (up to 106.7MPa) than either pure epoxy (83.8 MPa) or epoxy loaded with ZnO nanoparticles (91.5 MPa) [85].

7.3.2 Thermal and Chemical Properties

Metal-oxide nanoparticles with polymer matrix improve chemical and thermal stability due to their exceptional characteristics. Solid epoxy resin, DGEBA/biguanidine matrix, with metal oxides nanocomposites, were synthesized using a twin-screw extruder and two different nanoparticles based on alumina and zinc oxide at a 3 wt% composition [86]. After reinforcement, the glass transition temperature rises from 368 K in the pristine polymer to 377 K. Due to its short chain lengths, which may have increased reticulation close to the surface particles, the nanocomposite made of alumina has an activation energy of

61 kJ/mol [86]. By incorporating ZnO nanoparticles, Spirkova et al. have synthesized a variety of thermoplastic polycarbonate-based polyurethane (PC-PU) nanocomposites (0.5, 1, and 2 wt%). By using thermogravimetry (TGA) and differential scanning calorimetry (DSC) analyses, the thermal stability and other thermal properties of the produced nanocomposites were investigated [87]. The electrochemical oxidation of zinc metal was also used to synthesize a ZnO-chromophore nanomaterial, and thermally stable metal-organic chromophores have been produced. The synthesized nanocomposite exhibits 300°C thermal stability. When compared to metal-chromophore nanomaterials or pure chromophores, the photostability of the nanocomposites increased dramatically [88].

7.3.3 Electrical and Optical Properties

Researchers have long been interested in polymer/metal-oxide nanocomposites because of their electrical and optical properties. Since they need environmental protection, which is simple to incorporate into nanocomposite optical materials which are extremely difficult to originate from a single crystal. The polymer matrix is employed to hold the nanoparticles together, and the optical materials demonstrate improved characteristics at the nanoscale level. When the nanoparticles are on the nanoscale, there are several reasons why the electrical characteristics of polymer nanocomposites change. The interparticle distance first decreases when a particle's size drops for the same volume of fraction due to quantum processes, which leads to percolation and a rise in electrical characteristics. ZnO nanoparticles had a greater influence on reducing resistivity when added than when added to microfilters, which may be because of the high interfacial resistance and large interfacial area [89].

7.3.4 Biological Properties

Polymeric nanocomposites based on metal-oxide nanomaterials have a lot of potential in the biomedical, agricultural, and environmental sciences. Molecular imaging can benefit from iron oxide nanoparticles, which are tested *in vitro* as well as in animal studies. By adapting the enzyme-linked immunosorbent test, a cellular magnetic-linked immunosorbent assay was created as an MRI application for clinical diagnosis [90]. In order to determine the location and migration of cell culture following a transplant or blood transfusion employed iron oxide nanoparticles in MRI [91]. It involves relabeling iron oxide magnetic particles, which creates possibilities for using labeling magnetic particles in biology and medicine. For fluorescein-labeled protein, metal-oxide nanoparticles coated with biotin and bipyridinium carboxylic acid are utilized [92]. These nanoparticles serve as drug delivery carriers to treat cancer cells, which helps to mitigate the side effects of traditional chemotherapy. According to the observation, the magnetic field gradient causes the accumulation of nanoparticles at the tumor site [93, 94].

7.4 Applications

Polymer/metal-oxide nanoparticles have a variety of characteristics that are important for the vast range of potential applications. The following list provides some of the potential uses for polymer/metal-oxide hybrid nanocomposites (Table 7.1).

Table 7.1 Application of polymer/metal-oxide composites [95–133]

S. No.	Nanocomposites	Applications	References
1.	SiO_2-Nafion	DMFC membrane	Kim et al. [95]
2.	SiO_2-PS/PET	Engineering plastic reinforcement	Wu et al. [96]
3.	SiO_2-Epoxy	Coatings, moldings compounds	Kang et al. [97]
4.	SiO_2-Epoxy	adhesives, electronic packaging	Guo et al. [98]
5.	SiO_2-PMMA	Bioactive bone substitute, dental applications	Rhee et al. [99]
6.	SiO_2-PEG	Bioimaging	Maurice et al. [100]
7.	SiO_2-HPBBEA-g-PDMAEMA	Biomedicine	Li et al. [101]
8.	SiO_2-LBL-PEG	Controlled drug release	Du et al. [102]

(*Continued*)

Table 7.1 (*Continued*)

S. No.	Nanocomposites	Applications	References
9.	SnO_2-Nafion	PEMFC membrane	Scipioni et al. [103]
10.	WO_3-PC	Photochromic glass, building applications	Sangpraserdsuk et al. [104]
11.	ZrO_2-PVC	Engineering plastics	Mallakpour et al. [105]
12.	ZrO_2-Unsat. PE	Packaging applications	Adhikari et al. [106]
13.	ZrO_2 Cyanate ester/azomethine	Dielectric applications	Ariraman et al. [107]
14.	Al_2O_3-Unsat. PE	Packaging applications	Adhikari et al. [106]
15.	Al_2O_3-PC	Optical applications	Chandra et al. [108]
16.	Al_2O_3-PMMA	Dental and high-impact glasses	Siegel et al. [109]
17.	Al_2O_3-PMMA	Flame retardant materials	Cinausero et al. [110]
18.	MgO-LDPE	Electrical insulator	Ju et al. [111] Pourrahimi et al. [112]
19.	MgO-PLA	Biomedical materials, textiles, packaging materials	Li et al. [113]
20.	Fe_3O_4-POA, PVP	Magnetic applications	Dong et al. [114]
21.	Fe_3O_4-PMMA	Magnetic and electronic applications	Bach et al. [115]
22.	Fe_3O_4-PEG, PLA, PHEMA, PNIPAm	Biomedical and separation applications	Lattuada et al. [116]
23.	Fe_3O_4-SEBS	MW communication devices	Yang et al. [117]

S. No.	Nanocomposites	Applications	References
24.	Fe_3O_4-PDMS	Electromagnetic circuit	Pisanello et al. [118]
25.	TiO_2-AN-St-Aterpolymer	Solar reflectance, environmental applications	Qi et al. [119]
26.	TiO_2-LDPE	Electrical insulator	Ma et al. [120]
27.	TiO_2-PTAI	Sensors	Ghaemy et al. [121]
28.	TiO_2-Thiolene	Dental restorative materials	Schechtel et al. [122]
29.	TiO_2-High density polyethylene	Bone repair	Hashimoto et al. [123]
30.	TiO_2-PU	Self-cleaning and antibacterial coatings	Charpentier et al. [124]
31.	TiO_2-Epoxy resin	Flame retardants nanocomposites	Wu et al. [125]
32.	ZnO-PVC-EDTA	Biomedical devices	Mallakpour et al. [126]
33.	ZnO-PCL	Tissue engineering, food packaging	Mallakpour et al. [127]
34.	ZnO-PMMA-PS	UV shielding coating	Zeng et al. [128]
35.	ZnO-PA	Optical applications	Hajibeygi et al. [129]
36.	ZnO-LDPE	Electrical insulator	Pourrahimi et al. [130]
37.	ZnO-PS	Antistatic materials, UV shielding in coatings, films, and cosmetics	Ma et al. [131]
38.	ZnO-PHEA	UV absorber	Liu et al. [132]
39.	ZnO-SEBS and ZnO-SEBS-MA	HV insulation	Helal et al. [133], [134, 135]

7.5 Conclusions and Future Trends

This chapter mainly focuses on synthesis methods for producing polymer/metal-oxide nanocomposites using various physical and chemical processes and properties with applications. However, complete control over the size and form distribution of metal-oxide nanoparticles is still difficult to achieve. The crucial concerns that require our attention are coming up with less expensive, environmentally friendly ways to produce these oxides and gaining more control over how to manipulate the geometry of the nanostructured metal oxides. An *in vitro*, low-temperature biomimetic synthesis strategy can be used to create next-generation nanoengineered materials. For better investigation, it is necessary to thoroughly investigate the various properties of polymer/metal-oxide nanocomposites. It is beneficial to improve the functional properties of the polymers by loading them with extremely few metal-oxide nanoparticles. The status of the metal-oxide nanoparticles dispersion in the polymer matrix and the creation of the microstructural distribution during nanocomposite processing are two aspects that have an impact on the characteristics of the materials. The synthesis of metal-oxide nanoparticles tailored to certain applications would be of special interest to researchers; nevertheless, the problem of improving the compatibility, environmentally friendliness, and long-term stability of the metal-oxide nanoparticles still exists.

References

1. Bayda, Samer, Muhammad Adeel, Tiziano Tuccinardi, Marco Cordani, and Flavio Rizzolio. The history of nanoscience and nanotechnology: From chemical–physical applications to nanomedicine. *Molecules*, 25(1), 2019, 112.
2. Gleiter, Herbert. Nanostructured materials. *Advanced Materials*, 4(7–8), 1992, 474–481.
3. Park, Jin-Sung, Jin Koo Kim, Jeong Hoo Hong, Jung Sang Cho, Seung-Keun Park, and Yun Chan Kang. Advances in the synthesis and design of nanostructured materials by aerosol spray processes for efficient energy storage. *Nanoscale*, 11(41), 2019, 19012–19057.
4. Hahn, Horst. Unique features and properties of nanostructured materials. *Advanced Engineering Materials*, 5(5), 2003, 277–284.
5. Alves, Annelise Kopp, Carlos P. Bergmann, and Felipe Amorim Berutti. *Novel Synthesis and Characterization of Nanostructured Materials*, Springer, 2013.

6. Prasanna, SRV Siva, K. Balaji, Shyam Pandey, and Sravendra Rana. Metal oxide-based nanomaterials and their polymer nanocomposites. In: *Nanomaterials and Polymer Nanocomposites*, Elsevier, 2019, pp. 123–144.
7. Opoku, Francis, Ephraim M. Kiarii, Penny P. Govender, and Messai Adenew Mamo. *Metal Oxide Polymer Nanocomposites in Water Treatments*, Vol. 8, IntechOpen, London (UK), 2017.
8. Xu, Ping, Xijiang Han, Bin Zhang, Yunchen Du, and Hsing-Lin Wang. Multifunctional polymer–metal nanocomposites via direct chemical reduction by conjugated polymers. *Chemical Society Reviews*, 43(5), 2014, 1349–1360.
9. Shikuku, Victor Odhiambo, and Wilfrida N. Nyairo. Preparation and application of polymer-metal oxide nanocomposites in wastewater treatment: Challenges and potentialities. In: Clarizia, G. and P. Bernardo (eds.), *Diverse Applications of Organic-Inorganic Nanocomposites: Emerging Research and Opportunities*, IGI Global, 2020, pp. 83–102.
10. Zhao, Guixia, Xiubing Huang, Zhenwu Tang, Qifei Huang, Fenglei Niu, and Xiangke Wang. Polymer-based nanocomposites for heavy metal ions removal from aqueous solution: A review. *Polymer Chemistry*, 9(26), 2018, 3562–3582.
11. Camargo, Pedro Henrique Cury, Kestur Gundappa Satyanarayana, and Fernando Wypych. Nanocomposites: Synthesis, structure, properties and new application opportunities. *Materials Research*, 12, 2009, 1–39.
12. Nam, K., Y. Tsutsumi, C. Yoshikawa, Y. Tanaka, R. Fukaya, T. Kimura, H. Kobayashi, T. Hanawa, and A. Kishida. Preparation of novel polymer-metal oxide nanocomposites with nanophase separated hierarchical structure. *Bulletin of Materials Science*, 34(7), 2011, 1289–1296.
13. Jadhav, Niteen, Subramanyam Kasisomayajula, and Victoria Johnston Gelling. Polypyrrole/metal oxides-based composites/nanocomposites for corrosion protection. *Frontiers in Materials*, 7, 2020, 95.
14. Shameem, M. Muhammed, S. M. Sasikanth, Raja Annamalai, and R. Ganapathi Raman. A brief review on polymer nanocomposites and its applications. *Materials Today: Proceedings*, 45, 2021, 2536–2539.
15. Ahmed, Bilal, Animesh K. Ojha, Ajeet Singh, Florian Hirsch, Ingo Fischer, Donfack Patrice, and Arnulf Materny. Well-controlled in-situ growth of 2D WO_3 rectangular sheets on reduced graphene oxide with strong photocatalytic and antibacterial properties. *Journal of Hazardous Materials*, 347, 2018, 266–278.
16. Ahmed, Bilal, Sumeet Kumar, Animesh K. Ojha, P. Donfack, and A. Materny. Facile and controlled synthesis of aligned WO_3 nanorods and

nanosheets as an efficient photocatalyst material. *Spectrochimica Acta Part A: Molecular and Biomolecular Spectroscopy*, 175, 2017, 250–261.

17. Kumar, Sumeet, Bilal Ahmed, Animesh K. Ojha, Jayanta Das, and Ashok Kumar. Facile synthesis of CdO nanorods and exploiting its properties towards supercapacitor electrode materials and low power UV irradiation driven photocatalysis against methylene blue dye. *Materials Research Bulletin*, 90, 2017, 224–231.
18. Ahmed, Bilal, Animesh K. Ojha, Florian Hirsch, Ingo Fischer, Donfack Patrice, and Arnulf Materny. Tailoring of enhanced interfacial polarization in WO_3 nanorods grown over reduced graphene oxide synthesized by a one-step hydrothermal method. *RSC Advances*, 7(23), 2017, 13985–13996.
19. Singh, Arvind, Bilal Ahmed, Ajeet Singh, and Animesh K. Ojha. Photodegradation of phenanthrene catalyzed by rGO sheets and disk like structures synthesized using sugar cane juice as a reducing agent. *Spectrochimica Acta Part A: Molecular and Biomolecular Spectroscopy*, 204, 2018, 603–610.
20. Kumar, Sachin, Bilal Ahmed, Arvind Singh, Ajeet Singh, and Animesh K. Ojha. Experimental and theoretical investigations of unusual enhancement of room temperature ferromagnetism in nickel-cobalt codoped CeO_2 nanostructures. *Journal of Magnetism and Magnetic Materials*, 465, 2018, 756–761.
21. Singh, Arvind, Bilal Ahmed, Saurav K. Ojha, and Animesh K. Ojha. Role of annealing temperature on structural modification of MoO_3 for enhanced electrochemical properties. In: *Recent Research Trends in Energy Storage Devices*, Springer, Singapore, 2021, pp. 19–26.
22. Wahab, Rizwan, I. H. Hwang, Hyung-Shik Shin, Young-Soon Kim, Javed Musarrat, Abdulaziz A. Al-Khedhairy, and Maqsood A. Siddiqui. Zinc oxide nanostructures and their applications. In: Tiwari, A., A. K. Mishra, H. Kobayashi, A. P. F., Turner (eds.). *Intelligent Nanomaterials: Processes, Properties, and Applications*, 28, 2012, 183–212.
23. Hahn, Yoon-Bong. Zinc oxide nanostructures and their applications. *Korean Journal of Chemical Engineering*, 28(9), 2011, 1797–1813.
24. Liu, Xueqin, James Iocozzia, Yang Wang, Xun Cui, Yihuang Chen, Shiqiang Zhao, Zhen Li, and Zhiqun Lin. Noble metal–metal oxide nanohybrids with tailored nanostructures for efficient solar energy conversion, photocatalysis and environmental remediation. *Energy and Environmental Science*, 10(2), 2017, 402–434.

25. John, Riya Alice B., and A. Ruban Kumar. A review on resistive-based gas sensors for the detection of volatile organic compounds using metal-oxide nanostructures. *Inorganic Chemistry Communications*, 133, 2021, 108893.
26. Chen, Di, Qiufan Wang, Rongming Wang, and Guozhen Shen. Ternary oxide nanostructured materials for supercapacitors: A review. *Journal of Materials Chemistry A*, 3(19), 2015, 10158–10173.
27. Ikram, Muhammad, Mahak Rashid, Ali Haider, Sadia Naz, Junaid Haider, Ali Raza, M. T. Ansar et al. A review of photocatalytic characterization, and environmental cleaning, of metal oxide nanostructured materials. *Sustainable Materials and Technologies*, 30, 2021, e00343.
28. Nunes, Daniela, Ana Pimentel, Lidia Santos, Pedro Barquinha, Luis Pereira, Elvira Fortunato, and Rodrigo Martins. *Metal Oxide Nanostructures: Synthesis, Properties and Applications*, Elsevier, 2018.
29. Hassan, Israr U., Hiba Salim, Gowhar A. Naikoo, Tasbiha Awan, Riyaz A. Dar, Fareeha Arshad, Mohammed A. Tabidi, et al. A review on recent advances in hierarchically porous metal and metal oxide nanostructures as electrode materials for supercapacitors and non-enzymatic glucose sensors. *Journal of Saudi Chemical Society*, 25(5), 2021, 101228.
30. Periyasamy, Muthaimanoj, and Arik Kar. Modulating the properties of SnO_2 nanocrystals: Morphological effects on structural, photoluminescence, photocatalytic, electrochemical and gas sensing properties. *Journal of Materials Chemistry C*, 8(14), 2020, 4604–4635.
31. Kandasamy, Manikandan, Surjit Sahoo, Saroj Kumar Nayak, Brahmananda Chakraborty, and Chandra Sekhar Rout. Recent advances in engineered metal oxide nanostructures for supercapacitor applications: Experimental and theoretical aspects. *Journal of Materials Chemistry A*, 9(33), 2021, 17643–17700.
32. Wu, Junqiao, Jinbo Cao, Wei-Qiang Han, Anderson Janotti, and Ho-Cheol Kim (eds.). *Functional Metal Oxide Nanostructures*, Vol. 149, Springer Science and Business Media, 2011.
33. Kumar, Sunil, Vladimir Pavelyev, Prabhash Mishra, Nishant Tripathi, Prachi Sharma, and Fernando Calle. A review on 2D transition metal di-chalcogenides and metal oxide nanostructures based NO_2 gas sensors. *Materials Science in Semiconductor Processing*, 107, 2020, 104865.
34. Ashik, U. P. M., Shinji Kudo, and Jun Ichiro Hayashi. An overview of metal oxide nanostructures. *Synthesis of Inorganic Nanomaterials*, 2018, 19–57.

35. Song, Yonghai, Xia Li, Lanlan Sun, and Li Wang. Metal/metal oxide nanostructures derived from metal–organic frameworks. *RSC Advances*, 5(10), 2015, 7267–7279.

36. Tian, Wei, Hao Lu, and Liang Li. Nanoscale ultraviolet photodetectors based on one dimensional metal oxide nanostructures. *Nano Research*, 8(2), 2015, 382–405.

37. Galstyan, Vardan, Manohar P. Bhandari, Veronica Sberveglieri, Giorgio Sberveglieri, and Elisabetta Comini. Metal oxide nanostructures in food applications: Quality control and packaging. *Chemosensors*, 6(2), 2018, 16.

38. Joshi, Nirav, Takeshi Hayasaka, Yumeng Liu, Huiliang Liu, Osvaldo N. Oliveira, and Liwei Lin. A review on chemiresistive room temperature gas sensors based on metal oxide nanostructures, graphene and 2D transition metal dichalcogenides. *Microchimica Acta*, 185(4), 2018, 1–16.

39. Mirzaei, A., and G. Neri. Microwave-assisted synthesis of metal oxide nanostructures for gas sensing application: A review. *Sensors and Actuators B: Chemical*, 237, 2016, 749–775.

40. Zhai, Tianyou, Xiaosheng Fang, Meiyong Liao, Xijin Xu, Haibo Zeng, Bando Yoshio, and Dmitri Golberg. A comprehensive review of one-dimensional metal-oxide nanostructure photodetectors. *Sensors*, 9(8), 2009, 6504–6529.

41. Devan, Rupesh S., Ranjit A. Patil, Jin-Han Lin, and Yuan-Ron Ma. One-dimensional metal-oxide nanostructures: Recent developments in synthesis, characterization, and applications. *Advanced Functional Materials*, 22(16), 2012, 3326–3370.

42. Chen, Po-Chiang, Guozhen Shen, and Chongwu Zhou. Chemical sensors and electronic noses based on 1-D metal oxide nanostructures. *IEEE Transactions on Nanotechnology*, 7(6), 2008, 668–682.

43. Haldorai, Yuvaraj, and Jae-Jin Shim. Fabrication of metal oxide–polymer hybrid nanocomposites. *Organic-Inorganic Hybrid Nanomaterials*, 2014, 249–281.

44. Mallakpour, Shadpour, Zeinab Radfar, and Chaudhery Mustansar Hussain. Current advances on polymer-layered double hydroxides/metal oxides nanocomposites and bionanocomposites: Fabrications and applications in the textile industry and nanofibers. *Applied Clay Science*, 206, 2021, 106054.

45. Pandey, Bhamini, Poonam Singh, and Vinod Kumar. Photocatalytic-sorption processes for the removal of pollutants from wastewater using polymer metal oxide nanocomposites and associated

environmental risks. *Environmental Nanotechnology, Monitoring and Management*, 16, 2021, 100596.

46. Prasanna, S. R. V. Siva, K. Balaji, Shyam Pandey, and Sravendra Rana. Metal oxide based nanomaterials and their polymer nanocomposites. In: Karak, N. (ed.), *Nanomaterials and Polymer Nanocomposites: Raw Materials to Applications*, Elsevier, 2019, pp. 123–144.
47. Hashim, Ahmed, and Aseel Hadi. Novel pressure sensors made from nanocomposites (biodegradable polymers–metal oxide nanoparticles): Fabrication and characterization. *Ukrainian Journal of Physics* 63(8), 2018, 754–754.
48. Soytaş, Serap Hayat, Oğuzhan Oğuz, and Yusuf Ziya Menceloğlu. Polymer nanocomposites with decorated metal oxides. In: Pielichowski, K., and T. M. Majka (eds.), *Polymer Composites with Functionalized Nanoparticles*, Elsevier, 2019, pp. 287–323.
49. Shifrina, Zinaida B., Valentina G. Matveeva, and Lyudmila M. Bronstein. Role of polymer structures in catalysis by transition metal and metal oxide nanoparticle composites. *Chemical Reviews*, 120(2), 2019, 1350–1396.
50. Dakshayini, B. S., Kakarla Raghava Reddy, Amit Mishra, Nagaraj P. Shetti, Shweta J. Malode, Soumen Basu, S. Naveen, and Anjanapura V. Raghu. Role of conducting polymer and metal oxide-based hybrids for applications in ampereometric sensors and biosensors. *Microchemical Journal*, 147, 2019, 7–24.
51. Pourrahimi, Amir Masoud, Richard T. Olsson, and Mikael S. Hedenqvist. The role of interfaces in polyethylene/metal-oxide nanocomposites for ultrahigh-voltage insulating materials. *Advanced Materials*, 30(4), 2018, 1703624.
52. Shameem, M. Muhammed, S. M. Sasikanth, Raja Annamalai, and R. Ganapathi Raman. A brief review on polymer nanocomposites and its applications. *Materials Today: Proceedings*, 45, 2021, 2536–2539.
53. Shahidi, Sheila, Bahareh Moazzenchi, and Mahmood Ghoranneviss. A review-application of physical vapor deposition (PVD) and related methods in the textile industry. *The European Physical Journal Applied Physics*, 71(3), 2015, 31302.
54. Muratore, Christopher, Andrey A. Voevodin, and Nicholas R. Glavin. Physical vapor deposition of 2D van der Waals materials: A review. *Thin Solid Films*, 688, 2019, 137500.
55. Dai, Zu Rong, Z. Wei Pan, and Zhong L. Wang. Novel nanostructures of functional oxides synthesized by thermal evaporation. *Advanced Functional Materials*, 13(1), 2003, 9–24.

56. Mubarak, E. Hamzah, and M.R.M. Tofr. Review of physical vapor deposition (PVD) techniques for hard coating. *Jurnal Mekanikal*, 20, 2005, 42–51.

57. Campbell, S. A. *The Science and Engineering of Microelectronic Fabrication*. Oxford University Press, 2001, 585.

58. Krebs, Hans-Ulrich, Martin Weisheit, Jörg Faupel, Erik Süske, Thorsten Scharf, Christian Fuhse, Michael Störmer et al. Pulsed laser deposition (PLD): A versatile thin film technique. *Advances in Solid State Physics*, 2003, 505–518.

59. Kuppusami, P., and V. S. Raghunathan. Status of pulsed laser deposition: Challenges and opportunities. *Surface Engineering*, 22(2), 2006, 81–83.

60. Rella, R., J. Spadavecchia, M. G. Manera, S. Capone, A. Taurino, M. Martino, A. P. Caricato, and T. Tunno. Acetone and ethanol solid-state gas sensors based on TiO_2 nanoparticles thin film deposited by matrix assisted pulsed laser evaporation. *Sensors and Actuators B: Chemical*, 127(2), 2007, 426–431.

61. Sun, Luzhao, Guowen Yuan, Libo Gao, Jieun Yang, Manish Chhowalla, Meysam Heydari Gharahcheshmeh, Karen K. Gleason, Yong Seok Choi, Byung Hee Hong, and Zhongfan Liu. Chemical vapour deposition. *Nature Reviews Methods Primers*, 1(1), 2021, 1–20.

62. Ong, Chin Boon, Law Yong Ng, and Abdul Wahab Mohammad. A review of ZnO nanoparticles as solar photocatalysts, Synthesis, mechanisms and applications. *Renewable and Sustainable Energy Reviews*, 81, 2018, 536–551.

63. Polarz, Sebastian, Abhijit Roy, Michael Merz, Simon Halm, Detlef Schröder, Lars Schneider, Gerd Bacher, Frank E. Kruis, and Matthias Driess. Chemical vapor synthesis of size-selected zinc oxide nanoparticles. *Small*, 1(5), 2005, 540–552.

64. Amara, Daniel, Judith Grinblat, and Shlomo Margel. Solventless thermal decomposition of ferrocene as a new approach for one-step synthesis of magnetite nanocubes and nanospheres. *Journal of Materials Chemistry*, 22(5), 2012, 2188–2195.

65. Dehghanghadikolaei, Amir, Jamal Ansary, and Reza Ghoreishi. Sol-gel process applications: A mini-review. *Proceedings of the Nature Research Society*, 2(1), 2018, 02008–02029.

66. Livage, Jacques. Sol-gel processes. *Current Opinion in Solid State and Materials Science*, 2(2), 1997, 132–138.

67. Yilmaz, Erkan, and Mustafa Soylak. Functionalized nanomaterials for sample preparation methods. In: C. M. Hussain (ed.), *Handbook of Nanomaterials in Analytical Chemistry*, Elsevier, 2020, pp. 375–413.
68. Hench, Larry L., and Jon K. West. The sol-gel process. *Chemical Reviews*, 90(1), 1990, 33–72.
69. Hanif, Sara, and Asma Shahzad. Removal of chromium (VI) and dye Alizarin Red S (ARS) using polymer-coated iron oxide (Fe_3O_4) magnetic nanoparticles by co-precipitation method. *Journal of Nanoparticle Research*, 16(6), 2014, 1–15.
70. Dong, Hongxu, and Gary M. Koenig. A review on synthesis and engineering of crystal precursors produced via coprecipitation for multicomponent lithium-ion battery cathode materials. *CrystEngComm*, 22(9), 2020, 1514–1530.
71. Sarkar, Sudipta, Eric Guibal, Françoise Quignard, and A. K. SenGupta. Polymer-supported metals and metal oxide nanoparticles: Synthesis, characterization, and applications. *Journal of Nanoparticle Research*, 14(2), 2012, 1–24.
72. Nangai, E. Kayalvizhi and S. Saravanan. Synthesis, fabrication and testing of polymer nanocomposites: A review. *Materials Today: Proceedings*, 2021.
73. Campos, Eunice Aparecida, Denise Villela Barcza Stockler Pinto, José Irineu Sampaio de Oliveira, Elizabeth da Costa Mattos, and Rita de Cássia Lazzarini Dutra. Synthesis, characterization and applications of iron oxide nanoparticles - a short review. *Journal of Aerospace Technology and Management*, 7, 2015, 267–276.
74. Ye, Jianfeng, Wen Liu, Jinguang Cai, Shuai Chen, Xiaowei Zhao, Henghui Zhou, and Limin Qi. Nanoporous anatase TiO_2 mesocrystals: Additive-free synthesis, remarkable crystalline-phase stability, and improved lithium insertion behavior. *Journal of the American Chemical Society*, 133(4), 2011, 933–940.
75. M. Sh, A. A. M. Belal, and A. S. Al-Hussaini. From copolymer precursor to metal oxides nanoparticles: Synthesis and characterization of doped copper and cobalt copolymer via in situ and ex situ copolymerization. *Journal of Macromolecular Science, Part A*, 52(5), 2015, 394–400.
76. Schadler, Linda S. Polymer-based and polymer-filled nanocomposites. *Nanocomposite Science and Technology*, 2003, 77–153.

77. Sanchez, Clément, G. J. de A. A. Soler-Illia, François Ribot, T. Lalot, Cédric R. Mayer, and V. Cabuil. Designed hybrid organic–inorganic nanocomposites from functional nanobuilding blocks. *Chemistry of Materials*, 13(10), 2001, 3061–3083.

78. Erdem, Bedri, E. David Sudol, Victoria L. Dimonie, and Mohamed S. El-Aasser. Encapsulation of inorganic particles via miniemulsion polymerization. I. Dispersion of titanium dioxide particles in organic media using OLOA 370 as stabilizer. *Journal of Polymer Science Part A: Polymer Chemistry*, 38(24), 2000, 4419–4430.

79. Wu, Yanfei, Yang Zhang, Jiaxi Xu, Min Chen, and Limin Wu. One-step preparation of PS/TiO_2 nanocomposite particles via miniemulsion polymerization. *Journal of Colloid and Interface Science*, 343(1), 2010, 18–24.

80. Lü, Changli, and Bai Yang. High refractive index organic–inorganic nanocomposites: Design, synthesis and application. *Journal of Materials Chemistry*, 19(19), 2009, 2884–2901.

81. Miyauchi, Masahiro, Yongjin Li, and Hiroshi Shimizu. Enhanced degradation in nanocomposites of TiO_2 and biodegradable polymer. *Environmental Science and Technology*, 42(12), 2008, 4551–4554.

82. Dasari, Aravind, and James Njuguna (eds.). *Functional and Physical Properties of Polymer Nanocomposites*, John Wiley and Sons, 2016.

83. Mohan, S., J. Abraham, O. S., Oluwafemi, N., Kalarikkal, S., Thomas. Rheology and processing of inorganic nanomaterials and quantum dots/polymer nanocomposites. In: Thomas, S., R. Muller, and J. Abraham (eds.), *Rheology and Processing of Polymer Nanocomposites*, Wiley, 2016, pp. 355–382.

84. Maharramov, A. A., M. A. Ramazanov, Luca Di Palma, H. A. Shirinova, and F. V. Hajiyeva. Influence of magnetite nanoparticles on the dielectric properties of metal oxide/polymer nanocomposites based on polypropylene. *Russian Physics Journal*, 60(9), 2018, 1572–1576.

85. Liang, Chaobo, Ping Song, Hongbo Gu, Chao Ma, Yongqiang Guo, Hongyuan Zhang, Xiaojiang Xu, Qiuyu Zhang, and Junwei Gu. Nanopolydopamine coupled fluorescent nanozinc oxide reinforced epoxy nanocomposites. *Composites Part A: Applied Science and Manufacturing*, 102, 2017, 126–136.

86. Karasinski, E.N., Da Luz, M.G., Lepienski, C.M., Coelho, L.A.F., 2013. Nanostructured coating based on epoxy/metal oxides: Kinetic curing and mechanical properties. *Thermochimica Acta*, 569, 167–176.

87. Pavličević, Jelena, Milena Špírková, Oskar Bera, Mirjana Jovičić, Branka Pilić, Sebastian Baloš, and Jaroslava Budinski-Simendić. The influence of ZnO nanoparticles on thermal and mechanical behavior of polycarbonate-based polyurethane composites. *Composites Part B: Engineering*, 60, 2014, 673–679.

88. Skorenko, K., Bernier, R. T., Liu, J., Galusha, B., Goroleski, F., Hughes, B. P., Bernier, W. E., Jones, W. E. Thermal stability of ZnO nanoparticle bound organic chromophores. *Dyes and Pigments*, 131, 2016, 69–75.

89. Parala, Harish, Anjana Devi, Raghunandan Bhakta, and Roland A. Fischer. Synthesis of nano-scale TiO_2 particles by a nonhydrolytic approach. *Journal of Materials Chemistry*, 12(6), 2002, 1625–1627.

90. Burtea, C., S., Laurent, A., Roch, L., Vander Elst, R. N., Muller. C-MALISA (cellular magnetic-linked immunosorbent assay), a new application of cellular ELISA for MRI. *Journal of Inorganic Biochemistry*, 99, 2005, 1135–1144.

91. Bulte, J. W. M. Intracellular endosomal magnetic labeling of cells. In: Walker, J. M. (ed.), *Methods in Molecular Medicine*, 124, 2006, pp. 419–439

92. Fan, J., J. Lu, R. Xu, R. Jiang, and Y. Gao. Use of water-dispersible Fe_2O_3 nanoparticles with narrow size distributions in isolating avidin. *Journal of Colloid and Interface Science*, 266, 2003, 215–218.

93. Gallo, J. M., P. Varkonyi, E. E. Hassan, D. R. Groothius. Targeting anticancer drugs to the brain: II. Physiological pharmacokinetic model of oxantrazole following intraarterial administration to rat glioma-2 (RG-2) bearing rats. *Journal of Pharmacokinetics and Pharmacodynamics*, 21, 1993, 575–592.

94. Alexiou, C., Schmid, R.J., Jurgons, R., Kremer, M., Wanner, G., Bergemann, C., Huenges, E., Nawroth, T., Arnold, W., Parak, F.G., 2006. Targeting cancer cells: Magnetic nanoparticles as drug carriers. *European Biophysics Journal*, 35, 446–450.

95. Kim, Hyun-Jong, Yong-Gun Shul, and Haksoo Han. Sulfonic-functionalized heteropolyacid–silica nanoparticles for high temperature operation of a direct methanol fuel cell. *Journal of Power Sources*, 158(1), 2006, 137–142.

96. Wu, Tianbin, and Yangchuan Ke. Melting, crystallization and optical behaviors of poly (ethylene terephthalate)-silica/polystyrene nanocomposite films. *Thin Solid Films*, 515(13), 2007, 5220–5226.

97. Kang, Sungtack, Sung Il Hong, Chul Rim Choe, Min Park, Soonho Rim, and Junkyung Kim. Preparation and characterization of epoxy

composites filled with functionalized nanosilica particles obtained via sol–gel process. *Polymer*, 42(3), 2001, 879–887.

98. Guo, Qian, Pengli Zhu, Gang Li, Junjie Wen, Tianyu Wang, Daoqiang Daniel Lu, Rong Sun, and Chingping Wong. Study on the effects of interfacial interaction on the rheological and thermal performance of silica nanoparticles reinforced epoxy nanocomposites. *Composites Part B: Engineering*, 116, 2017, 388–397.

99. Rhee, Sang-Hoon, and Je-Yong Choi. Preparation of a bioactive poly (methyl methacrylate)/silica nanocomposite. *Journal of the American Ceramic Society*, 85(5), 2002, 1318–1320.

100. Maurice, Vincent, Cedric Slostowski, Nathalie Herlin-Boime, and Geraldine Carrot. Polymer-grafted silicon nanoparticles obtained either via peptide bonding or click chemistry. *Macromolecular Chemistry and Physics*, 213(23), 2012, 2498–2503.

101. Li, Xin, Chun-Yan Hong, and Cai-Yuan Pan. Preparation and characterization of hyperbranched polymer grafted mesoporous silica nanoparticles via self-condensing atom transfer radical vinyl polymerization. *Polymer*, 51(1), 2010, 92–99.

102. Du, Pengcheng, Xubo Zhao, Jin Zeng, Jinshan Guo, and Peng Liu. Layer-by-layer engineering fluorescent polyelectrolyte coated mesoporous silica nanoparticles as pH-sensitive nanocarriers for controlled release. *Applied Surface Science*, 345, 2015, 90–98.

103. Scipioni, Roberto, Delia Gazzoli, Francesca Teocoli, Oriele Palumbo, Annalisa Paolone, Neluta Ibris, Sergio Brutti, and Maria Assunta Navarra. Preparation and characterization of nanocomposite polymer membranes containing functionalized SnO_2 additives. *Membranes*, 4(1), 2014, 123–142.

104. Sangpraserdsuk, Tanes, Manisara Phiriyawirut, Pailin Ngaotrakanwiwat, and Jatuphorn Wootthikanokkhan. Mechanical, optical, and photochromic properties of polycarbonate composites reinforced with nano-tungsten trioxide particles. *Journal of Reinforced Plastics and Composites*, 36(16), 2017, 1168–1182.

105. Mallakpour, Shadpour, and Ahmadreza Nezamzadeh Ezhieh. A simple and environmentally friendly method for surface modification of ZrO_2 nanoparticles by biosafe citric acid as well as ascorbic acid (vitamin C) and its application for the preparation of poly (vinyl chloride) nanocomposite films. *Polymer Composites*, 38(8), 2017, 1756–1765.

106. Adhikari, Jaideep, Bhabatosh Biswas, Sumit Chabri, Nil Ratan Bandyapadhyay, Pravin Sawai, Bhairab Chandra Mitra, and Arijit Sinha. Effect of functionalized metal oxides addition on the

mechanical, thermal and swelling behaviour of polyester/jute composites. *Engineering Science and Technology, an International Journal*, 20(2), 2017, 760–774.

107. Ariraman, Mathivathanan, Ramachandran Sasi Kumar, and Muthukaruppan Alagar. Design of cyanate ester/azomethine/ZrO_2 nanocomposites high-k dielectric materials by single step sol–gel approach. *Journal of Applied Polymer Science*, 131(24), 2014.

108. Chandra, Alexander, Lih-Sheng Turng, Padma Gopalan, Roger M. Rowell, and Shaoqin Gong. Study of utilizing thin polymer surface coating on the nanoparticles for melt compounding of polycarbonate/alumina nanocomposites and their optical properties. *Composites Science and Technology*, 68(3–4), 2008, 768–776.

109. Siegel, R. W., S. K. Chang, B. J. Ash, J. A. P. M. Stone, P. M. Ajayan, R. W. Doremus, and L. S. Schadler. Mechanical behavior of polymer and ceramic matrix nanocomposites. *Scripta Materialia*, 44(8–9), 2001, 2061–2064.

110. Cinausero, Nicolas, Nathalie Azema, Marianne Cochez, Michel Ferriol, Mohamed Essahli, Francois Ganachaud, and José-Marie Lopez-Cuesta. Influence of the surface modification of alumina nanoparticles on the thermal stability and fire reaction of PMMA composites. *Polymers for Advanced Technologies*, 19(6), 2008, 701–709.

111. Ju, Siting, Hui Zhang, Mingji Chen, Chong Zhang, Xin Chen, and Zhong Zhang. Improved electrical insulating properties of LDPE based nanocomposite: Effect of surface modification of magnesia nanoparticles. *Composites Part A: Applied Science and Manufacturing*, 66, 2014, 183–192.

112. Pourrahimi, Amir Masoud, Love KH Pallon, Dongming Liu, Tuan Anh Hoang, Stanislaw Gubanski, Mikael S. Hedenqvist, Richard T. Olsson, and Ulf W. Gedde. Polyethylene nanocomposites for the next generation of ultralow-transmission-loss HVDC cables: Insulation containing moisture-resistant MgO nanoparticles. *ACS Applied Materials and Interfaces*, 8(23), 2016, 14824–14835.

113. Li, Yonghui, and Xiuzhi Susan Sun. Preparation and characterization of polymer–inorganic nanocomposites by in situ melt polycondensation of L-lactic acid and surface-hydroxylated MgO. *Biomacromolecules*, 11(7), 2010, 1847–1855.

114. Dong, Angang, Xingchen Ye, Jun Chen, Yijin Kang, Thomas Gordon, James M. Kikkawa, and Christopher B. Murray. A generalized ligand-exchange strategy enabling sequential surface functionalization of colloidal

nanocrystals. *Journal of the American Chemical Society*, 133(4), 2011, 998–1006.

115. Bach, Long Giang, Md Rafiqul Islam, Jong Tae Kim, SungYong Seo, and Kwon Taek Lim. Encapsulation of Fe_3O_4 magnetic nanoparticles with poly (methyl methacrylate) via surface functionalized thiol-lactam initiated radical polymerization. *Applied Surface Science*, 258(7), 2012, 2959–2966.

116. Lattuada, Marco, and T. Alan Hatton. Functionalization of monodisperse magnetic nanoparticles. *Langmuir*, 23(4), 2007, 2158–2168.

117. Yang, Ta-I., Rene NC Brown, Leo C. Kempel, and Peter Kofinas. Magneto-dielectric properties of polymer–Fe_3O_4 nanocomposites. *Journal of Magnetism and Magnetic Materials*, 320(21), 2008, 2714–2720.

118. Pisanello, Ferruccio, Rosa De Paolis, Daniela Lorenzo, Simone Nitti, Giuseppina Monti, Despina Fragouli, Athanassia Athanassiou et al. Radiofrequency characterization of polydimethylsiloxane–iron oxide based nanocomposites. *Microelectronic Engineering*, 111, 2013, 46–51.

119. Qi, Yanli, Bo Xiang, Wubin Tan, and Jun Zhang. Hydrophobic surface modification of TiO_2 nanoparticles for production of acrylonitrile-styrene-acrylate terpolymer/TiO_2 composited cool materials. *Applied Surface Science*, 419, 2017, 213–223.

120. Ma, Dongling, Treese A. Hugener, Richard W. Siegel, Anna Christerson, Eva Mårtensson, Carina Önneby, and Linda S. Schadler. Influence of nanoparticle surface modification on the electrical behaviour of polyethylene nanocomposites. *Nanotechnology*, 16(6), 2005, 724.

121. Ghaemy, Mousa, Soudabe Qasemi, Khadijeh Ghassemi, and Maasoomeh Bazzar. Nanostructured composites of poly (triazole-amide-imide)s and reactive titanium oxide by epoxide functionalization: Thermal, mechanical, photophysical and metal ions adsorption properties. *Journal of Polymer Research*, 20(10), 2013, 1–15.

122. Schechtel, Eugen, Yaping Yan, Xiangfan Xu, Yu Cang, Wolfgang Tremel, Zuyuan Wang, Baowen Li, and George Fytas. Elastic modulus and thermal conductivity of thiolene/TiO_2 nanocomposites. *The Journal of Physical Chemistry C*, 121(45), 2017, 25568–25575.

123. Hashimoto, Masami, Hiroaki Takadama, Mineo Mizuno, and Tadashi Kokubo. Enhancement of mechanical strength of TiO_2/high-density polyethylene composites for bone repair with silane-coupling treatment. *Materials Research Bulletin*, 41(3), 2006, 515–524.

124. Charpentier, P. A., K. Burgess, L. Wang, R. R. Chowdhury, A. F. Lotus, and G. Moula. Nano-TiO_2/polyurethane composites for antibacterial and self-cleaning coatings. *Nanotechnology*, 23(42), 2012, 425606.

125. Wu, Yu, Lei Song, and Yuan Hu. Fabrication and characterization of TiO_2 nanotube–epoxy nanocomposites. *Industrial and Engineering Chemistry Research*, 50(21), 2011, 11988–11995.

126. Mallakpour, Shadpour, and Mashal Javadpour. Antimicrobial, mechanical, optical and thermal properties of PVC/ZnO-EDTA nanocomposite films. *Polymers for Advanced Technologies*, 28(3), 2017, 393–403.

127. Mallakpour, Shadpour, and Nasrin Nouruzi. Effect of modified ZnO nanoparticles with biosafe molecule on the morphology and physiochemical properties of novel polycaprolactone nanocomposites. *Polymer*, 89, 2016, 94–101.

128. Zeng, Xiao-Fei, Xiang-Rong Kong, Jun-Lin Ge, Hai-Tao Liu, Cui Gao, Zhi-Gang Shen, and Jian-Feng Chen. Effective solution mixing method to fabricate highly transparent and optical functional organic–inorganic nanocomposite film. *Industrial and Engineering Chemistry Research*, 50(6), 2011, 3253–3258.

129. Hajibeygi, Mohsen, Meisam Shabanian, Mehrdad Omidi-Ghallemohamadi, and Hossein Ali Khonakdar. Optical, thermal and combustion properties of self-colored polyamide nanocomposites reinforced with azo dye surface modified ZnO nanoparticles. *Applied Surface Science*, 416, 2017, 628–638.

130. Pourrahimi, Amir M., Tuan A. Hoang, Dongming Liu, Love KH Pallon, Stanislaw Gubanski, Richard T. Olsson, Ulf W. Gedde, and Mikael S. Hedenqvist. Highly efficient interfaces in nanocomposites based on polyethylene and ZnO nano/hierarchical particles: A novel approach toward ultralow electrical conductivity insulations. *Advanced Materials*, 28(39), 2016, 8651–8657.

131. Ma, Chen-Chi M., Yi-Jie Chen, and Hsu-Chiang Kuan. Polystyrene nanocomposite materials—preparation, mechanical, electrical and thermal properties, and morphology. *Journal of Applied Polymer Science*, 100(1), 2006, 508–515.

132. Liu, Peng, and Tingmei Wang. Poly (hydroethyl acrylate) grafted from ZnO nanoparticles via surface-initiated atom transfer radical polymerization. *Current Applied Physics*, 8(1), 2008, 66–70.

133. Helal, Emna, Eric David, Michel Frechette, and Nicole R. Demarquette. Thermoplastic elastomer nanocomposites with

controlled nanoparticles dispersion for HV insulation systems: Correlation between rheological, thermal, electrical and dielectric properties. *European Polymer Journal*, 94, 2017, 68–86.

134. Yusuf, M. and M. Shahid (eds.). *Emerging Technologies for Textile Coloration*, CRC Press, Boca Raton, 2022.

135. Yusuf, M. and A. Madhu. Smart nanotextiles for filtration. In: Yilmaz, N. D. (ed.), *Smart Nanotextiles: Wearable and Technical Applications*, Scrivener Publishing, USA, 2022, pp. 341–360.

Chapter 8

Self-Cleaning Nanofinishes and Applications

Akhtarul Islam Amjad
Department of Fashion Technology, National Institute of Fashion Technology, Panchkula, Haryana, India
akhtarul.amjad@nift.ac.in

The realization of the self-cleaning and self-sanitizing concept on the textile substrate by nanotechnology has piqued the imagination of many people due to its unique characteristics and versatility. The self-cleaning fabrics are manufactured using fluorocarbons or nanotechnology. Due to less durability and the presence of fluorine contained in the fluorocarbon finishing, nanotechnology is being accepted at a very fast pace. Photocatalysts, carbon nanotubes, silver nanoparticles, chlorine halamine, and metal oxide colloidal are applied to develop self-cleaning fabrics. Water, oil repellency, and antibacterial potency are the wider features of today's self-cleaning finishes. This chapter reviews such important areas of self-cleaning finishing, mainly focusing on different nanomaterials, technology, and application of self-cleaning fabric.

Handbook of Nanofibers and Nanocomposites: Characteristics, Synthesis, and Applications in Textiles
Edited by Mohd Yusuf and Aminoddin Haji

ISBN 978-981-4968-77-5 (Hardcover), 978-1-003-43274-6 (eBook)
www.jennystanford.com

8.1 Introduction

Increased customer demand for durable and practical clothing over the last few decades has given rise to the possibility of nanoparticles being adhered to textile substrates as an active functionalizing agent to change their surface properties [1]. Nanomaterials are important in technological evolution because of their unique surface qualities, which allow them to have a greater impact than bulky standard additives and materials [2]. Nanostructured materials are utilized in antibacterial, anti-odor, ultra violates protection, stain repellency, self-clear, and other functional treatments in the form of metal-oxide agents, carbon-based materials, halamine compounds, etc. [1, 2]. Self-cleaning surfaces are substances that have an inbuilt ability to remove impurities such as soil, hazardous chemicals, and bacteria from their surface in a number of ways [3]. Nature is the basis of inspiration and motivation for self-cleaning technologies. Several natural surfaces have self-cleaning qualities. Butterfly wings and plant leaves such as cabbage and lotus are examples [4]. Due to the diversified applications, such as outdoor glass cleaning, solar panel cleaning, cement, and textiles, self-cleaning technology drew a lot of attraction in the late twentieth century. There are several projects underway throughout the globe to produce highly efficient, robust, and long-lasting self-cleaning coating surfaces with improved abilities [2–5]. Instead of a wider range of applications, self-cleaning technology also provides low maintenance costs, the exclusion of laborious work, and a decrease in the time spent on cleaning work. Textiles and apparel, in general, become dirty after everyday use, necessitating frequent cleaning with soap and water, resulting in time, money, and water waste [6].

Self-cleaning fabrics are those that can clean themselves without needing to be laundered [2]. Hydrophilic and hydrophobic are the two basic classifications for self-cleaning finishes or coatings. A hydrophilic coating causes the spreading of water (sheeting of water) across the surfaces, carrying filth, contamination, and other impurities away. In contrast, a hydrophobic coating leads to the sliding and rolling of the water drops across the surfaces and cleaning them [3, 4, 7]. However, hydrophilic coatings containing suitable metal oxides benefit from chemically breaking down complicated dirt deposits via a sunlight-assisted cleaning mechanism [8]. This characteristic can be obtained in textiles by making water and soil-repellent textile surface [6–8]. The surface energy and contact angle are directly related to water and oil repellency [9]. Coatings or finishes generate a distinct surface on the fabric, and variations in the microscopic properties of the surface govern the self-cleaning ability [10]. Several chemical solutions and nanotechnology are used to improve the coatings for optimal surface roughness or decrease surface energy [7–11]. The advancement of several processes such as plasma technology, sol-gel method, electrospraying, chemical etching, electrospinning, and nanoparticles/chemical synthesis has paved the robust and fast way for self-cleaning fabrics [12]. Researchers have developed self-cleaning fabrics from almost all textile materials [7, 8, 13–15].

This chapter summarizes a set of materials, mechanisms for forming self-cleaning coatings, future applications, and challenges. All past efforts to make self-cleaning textiles by increasing surface roughness or decreasing surface energy utilizing low-surface-energy chemicals or nanotechnology are included. Their fundamental working parts have also been thoroughly addressed.

8.2 Inspiration for Self-Cleaning Surface and Historical Development

Researchers have been working on self-cleaning surfaces since the 1960s [9–11, 16]. In 1995, the first commercial self-cleaning surface was invented in the form of a transparent layer of titanium dioxide (TiO_2) coating on the glass to allow self-clean [17]. In 2001, Pilkington glass was also one of the major initial milestones of self-cleaning surfaces [18]. A two-stage cleaning technique is used in this product. The photocatalysis of any fouling particles on the glass is the first stage. Following this stage, the glass becomes super-hydrophilic, allowing water to wash away the catalyzed residues on the surface. Titanium dioxide has been utilized to make self-cleaning nanoparticles that can be incorporated into various material surfaces since the invention of self-cleaning glass [11, 17, 18].

Many researchers throughout the world are inspired by nature to produce attractive self-cleaning functional systems or materials

[19–21]. In Asian religions, the lotus blossom is regarded as a sign of purity. The self-cleaning idea was inspired by natural phenomena such as lotus plant leaves, rice plant leaves, butterfly wings, fish scales, and so on [4]. Lotus is of plant that can thrive in the mud without the mud affecting the plant's purity. The lubricated surfaces of lotus leaves, along with the existence of tiny structures, result from an exceptionally hydrophobic surface [19, 20]. Barthlott and Neihuis [20] investigated the ability of plant leaves to self-clean their surfaces in 1997. In the mid-1960s, a high-resolution scanning electron microscope was used to examine the surface feature. According to SEM studies, surfaces that seem to be macroscopically even and plane have microscopic roughness on varying scale lengths [9, 10]. The leaves are extremely hydrophobic due to their surfaces and the presence of epicuticular wax crystalloids. These discoveries paved the door for developing a variety of self-cleaning ultra-hydrophobic surfaces inspired by nature.

Researchers discovered two basic forms of surface in plant leaves with superhydrophobicity on the micro scales. According to their findings, one is ordered micro and nanostructures, and another is unitary micro-line structures [4, 19–21]. A common example of a self-cleaning capability among insect surfaces is the cicada wing. The s regularly aligned nano-posts of the cicada wing endow it the superhydrophobicity (WCA160°) and hence self-cleaning capabilities, allowing dust particles to be successfully removed from the wing surface [7, 19–21]. Apart from lotus leaves and cicada wings, the surfaces of some other plants and insects also show a superhydrophobic self-cleaning property, such as rice leaves, the feathers of ducks and geese, and butterfly wings. However, their surfaces exhibit anisotropic wettability, which is contrary to the surfaces of lotus leaves and cicada wings [20–22]. Butterfly wings are another example of anisotropic wettability and have been found superhydrophobic, with a 152 ± 2° water contact angle. A water droplet can readily slide off the surface of a butterfly wing in one direction while being tightly trapped oppositely [4, 19].

Rice leaves have anisotropic wettability. Rice leaves have a hierarchical micro-nanoscale structure comparable to lotus leaves on their surface. On the other hand, papillae are parallelly arranged in a well-ordered form to the leaf edges, while the vertical direction is disordered. Another example is a fish body; even though the sea is polluted by oil, the fish protects the body from plankton and have a self-cleaning property in the sea [21, 22]. Researchers discovered that fish scales had super-oleophobic characteristics in air and super-oleophobic characteristics in water, with a contact angle of 156.4 ± 3.0° for an oil droplet (1,2-dichloroethane) [4, 71–20].

8.3 Basic Mechanisms of Self-Cleaning

Researchers are continuously working on novel technologies and mechanisms to generate and regulate low surface energy on a micron and nanometer scale, inspired by nature's superhydrophobic qualities [5, 7, 10, 18, 19]. Controlling surface morphology brings up a world of options for creating a wide range of engineered surfaces [22–24]. If the surface tension of a solid is greater than the surface tension of the liquid, a liquid droplet can wet a solid surface. As a result, in order to achieve liquid repellency, the solid's surface tension must be lower than that of the liquid. Wettability is the ability to establish contact with a solid surface due to the intermolecular forces of a liquid. The quantity of wettability is determined by the balance of cohesion and adhesion. The wetness of a drop of liquid can be determined by the angle at which it makes contact with the surface [9, 25]. The contact angle is formed by a solid textile surface and a contact-fluid droplet. It is the angle produced between the liquid drop and solid surface at the three-phase boundary (solid/liquid/vapor). Figure 8.1 shows the mechanism of wetting at different contact angles. A solid surface is hydrophilic if the contact angle is lesser than 90°. The surface is defined as hydrophobic when the contact angle is greater than 90°. Similarly, a surface with a water contact angle near zero is considered ultra (super) hydrophilic, while a surface with a contact angle of more than 150° is considered ultra (super) hydrophobic. An extremely hydrophobic surface generates water or oil repellence, resulting in a self-cleaning activity [4, 25, 26].

Coating textile surfaces with superhydrophobic materials by controlling wettability (silicones, fluorocarbons, etc.) and coating with functional hydrophilic materials via nanotechnology or photocatalytic characteristics are the two main methods of self-cleaning materials [1, 7, 27]. However, hydrophilic coatings have an extra quality that allows them to break down the adsorbed dirt in sunlight using photocatalysts chemically [6, 7].

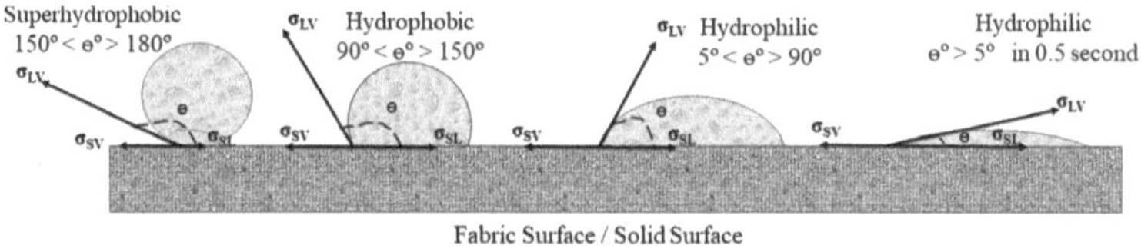

Figure 8.1 Conditions of hydrophobic, superhydrophobic, hydrophilic, and superhydrophilic surface [7, 21–26].

Photocatalytic self-cleaning is based on two separate mechanisms that work together to create a cleaning effect [17, 18, 28]. Dirt removal occurs as a result of the direct oxidation of pollutants, which is assisted by the effect of super hydrophilicity on catalytically active surfaces in contact with water (easy-to-clean effect). Due to their remarkable self-cleaning and antibacterial qualities, semiconductor materials such as titanium dioxide have sparked a lot of attention. These semiconductors photocatalytically produced superhydrophilic surfaces give excellent self-cleaning properties for industrial, commercial, and residential applications. UV or visible light irradiates these semiconductors during the photochemical reaction, accelerating the electrons to excited states and facilitating the production of reactive oxygen species (ROS). These ROS destroy organic molecules and cause super hydrophilicity in their targets or surface [28, 29].

8.4 Methodology for Developing the Self-Cleaning Textile Material

In the last few decades, various methods of water-, soil-, and oil-repellent finishes of fibers/fabrics have been of rising interest among textile technocrats and researchers because they impart characteristics such as reduced cleaning efforts and costs. Water-, oil-, and soil-resistant fabrics with self-cleaning characteristics can be made using a variety of materials and procedures. Surface treatments are used to obtain a situation of finite wettability or repellency, or hydrophilicity, which pave the thought for self-cleaning textiles [1, 8, 24].

8.4.1 Self-Cleaning Textile Finishing by a Hydrophobic Material

8.4.1.1 Plasma technology

Plasma treatment is an ecologically friendly method for permanently altering the surface of textiles [12]. This method is fiber-independent and can affect the top nanometers of a fiber's surface while maintaining the bulk properties. The demand to make the textile materials "self-clean" or at the very "easy-to-clean" is a significant common challenge. This can be done with the help of plasma technology. Plasma pretreatments can achieve self-cleaning effects in textile fabrics by increasing their water repellency to the fullest extent possible without altering their bulk qualities. Working temperatures, pressures, voltages, and frequencies can all be used to classify plasma treatment equipment [13]. Cleaning, activation, grafting, etching, and polymerization are some surface alterations that plasma therapy can make to fibers [13, 14, 30].

Plasma etching is the process of physically eradicating surface material from the substrate and forming volatile compounds through chemical reactions at the surface. Lower molecular weight fragments will be present on the etched surface. For etching polymers, inert gases (argon, helium, etc.), nitrogen or oxygen plasmas are commonly utilized [30]. Surface activation is the process of adding novel functional groups to a coated surface in order to change its surface energy and give it new capabilities. Plasma activation takes place in non-polymerizing gases. The bombardment of the treated material with reactive plasma species breaks covalent bonds and produces free radicals. These radicals react with active plasma species on the substrate surface. It creates active chemical functional groups such as hydroxyl, carbonyl, carboxyl, and amine groups [14, 31]. This type of activation changes the chemical activity and surface properties. Plasma polymerization creates a thin polymer layer on the substrate surface by polymerizing an organic monomer in plasma, such as methane (CH_4), ethane (C_2H_6), tetrafluoroethylene (C_2F_4), or hexafluoropropylene (C_3F_6). Plasma bombardment of the surface prepares the surface for reactive bonding sites. The plasma then mounts and grafts molecular fragments to the surface while polymerizing. Based on the precise gas and process conditions used, the deposited thin coatings can have a variety of qualities [13, 30–32].

It was discovered that modifying PET fibers with such a process significantly improves the self-cleaning properties such as antibacterial and color stain degradation [33]. This conclusion was attributable to the higher active sites and bondability on the plasma-treated polyester surface [31–33]. The increased self-cleaning properties of oxygen and argon RF plasma-modified polyester textiles with hydrophilic TiO_2 were also emphasized and investigated. Plasma activation of polyester fibers increased the loading of colloidal TiO_2 nanoparticles onto the surface, resulting in an outstanding photocatalytic activity. Furthermore, it was discovered that oxygen plasma pretreatment was preferable to argon plasma pretreatment [13, 34]. After RF plasma and TiO_2 treatments, cotton and nylon fabrics also show significant self-cleaning properties. This outcome was due to plasma-induced surface modification, which resulted in enhanced TiO_2 and textile contact [33–36].

8.4.1.2 Electrospinning technology

Electrospinning is introduced as a revolutionary engineering technology for creating superhydrophobic surfaces. The electrospinning technology has recently been widely used to create superhydrophobic electrospun fibers because the small fiber diameters contribute to super-hydrophobicity and results in unique surface roughness/texture [6]. Incorporating some nanoparticles into the post-treatment of electrospun fibers has also been used to make superhydrophobic nanofibers that contribute to extra roughness. To achieve this unique wettability, two critical elements must be controlled on at least two separate length scales: low surface energy and ordered surface roughness (i.e., micrometric and nanometric morphology) [37]. Figure 8.2 depicts the electrospinning technology. From Fig. 8.2, an electrostatic force is employed in the electrospinning process to create an electrically charged polymeric jet that controls the surface tension of the polymeric solution [37, 38]. As a result, elongated fibers are expedited from the capillary tip. These fibers are placed onto the collector with the matching solvent evaporation. During the manufacturing of fibers, reasonable monitoring of their morphological, optical, and wetting characteristics are required. With the help of electrospinning, ultrathin or nanofibers can be produced by strictly controlling many limits such as voltage, flow velocity, the viscosity of the polymeric precursor, and tip-to-tip collector distance [28, 37, 38]. Controlling these parameters well allows for the production of electrospun fibers with the required shape (mostly size, porosity, and fiber diameter). Because of the presence of fibers with micrometric and sub-micrometric diameters, surface modification to control the wettability of electrospun mats is conceivable [37–39]. A gradual rise in the research based on the combination of electrospun fibers and superhydrophobic surfaces were found in recent time. Potential uses of such technology are oil separation, filtration, conductivity generation, UV protection, and biological treatments [6, 28, 40].

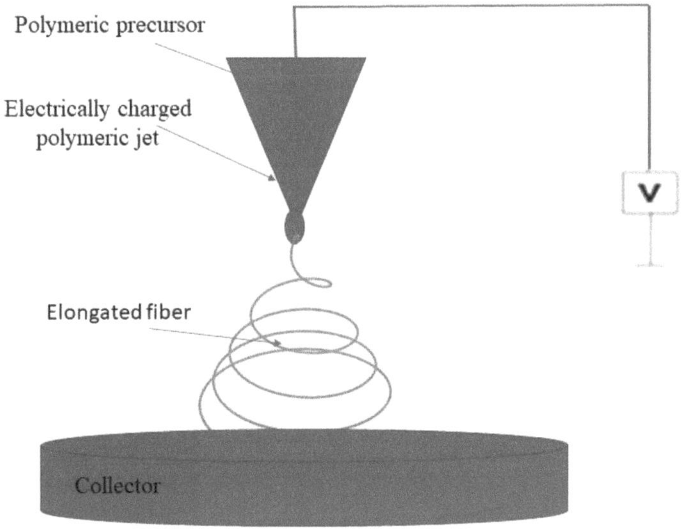

Figure 8.2 Electrospinning to develop self-cleaning surface [37–39].

8.4.1.3 Sol-gel treatment

Due to its unique features, such as low-temperature processing, functionalization of fiber surfaces, and high homogeneity of final products, the sol-gel technique has recently become widely applied for developing multifunction clothing [41]. The sol-gel process is an easy and low-cost way to make superhydrophobic coatings or films. It has a significant impact on the chemical finishing of textiles

[22, 42]. The initial task of the sol-gel process is the hydrolysis of a metal alkoxide (known as nucleation), and the second stage is a polycondensation (growth) reaction [43]. Figure 8.3 shows the sol-gel technique for developing self-cleaning fabric.

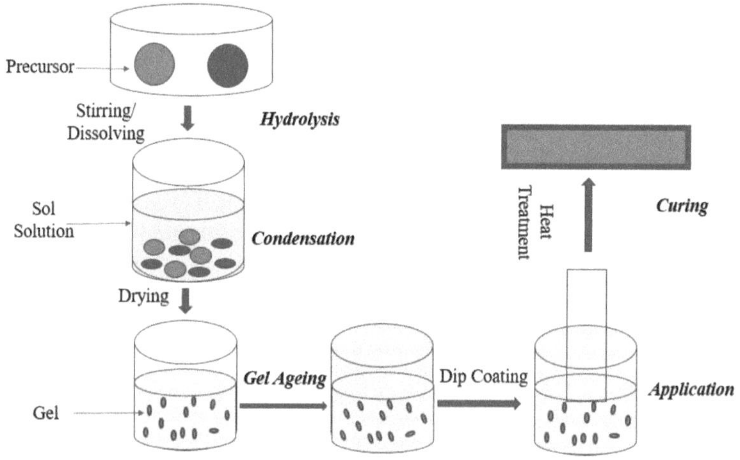

Figure 8.3 Sol-gel technique for developing self-cleaning fabric [42, 43, 48].

According to Fig. 8.3, in the first step, the precursor is dissolved in a solvent reaction. Water or an organic solvent is used regardless of whether the precursor is an inorganic salt or a metal oxide. Hydrolysis, condensation (gelation), gel aging, applications, and curing are the five phases of sol-gel processes. The sol is made by hydrolyzing or alcoholizing the solute or solvent, which results in small nanometer-sized particles. A non-ionized raw material, such as a metal alkoxide M(OR)n, interacts with water (R represents an alkyl group while M is metal with n valence).

The majority of work is based on the usage of silicon alkoxides as precursors in sol-gel production, such as tetraethoxysilane (TEOS) [41–43]. Using TEOS as a precursor has a number of advantages, including ease of size distribution control and good compatibility with other organic additives. In an organic matrix, silica produces a network of oxides. The hydrolysis of TEOS molecules [$Si(OC_2H_5)_4$] produces silanol groups during this process. Siloxane bridges (Si–O–Si) are formed by condensation or polymerization between silanol groups or between silanol groups and ethoxy groups, yielding the silica structure. The sol-gel process has a restriction in terms of washing durability of hydrophobic characteristics [42–44]. Multifunctional compounds, such as silane coupling agents, have been shown to improve adhesion and decrease the hydrophilicity of substrates, particularly cellulose fibers [43].

8.4.1.4 Fluorinated polymers

Because of their extraordinarily low surface energies, fluorinated polymers are gaining a lot of attention these days. Superhydrophobic surfaces can be created by roughening these polymers [46]. Fluorocarbons contain a perfluorinated carbon chain. It has a surface tension of less than 10 dyne/cm. Fluorocarbons form a thin coating surrounding the textile surface during the application procedure [47]. It leads to lower surface tension. As a result, coated textiles show water repellency. A water droplet then rolls away from the textile surface since it does not attach to it. Silicones are organosilicon compounds that have been extensively studied for use in the manufacture of ultra-hydrophobic textile surfaces. Researchers have developed for creating extremely hydrophobic self-cleaning surfaces with silicones. PDMS (polydimethylsiloxane) is an organosilicon chemical that is frequently referred to as silicone. Because of its inherent deformability and hydrophobic characteristics, PDMS is an excellent material for creating superhydrophobic surfaces. PDMS is used in various ways to create superhydrophobic surfaces [48]. PDMS (polydimethylsiloxane) can be utilized for surface alteration of textiles by stimulating a CO_2 pulsed laser. It introduces peroxide groups onto the PDMS surface, resulting in the roughness of the surface. These groups help 2-hydroxyethyl methacrylate (HEMA) graft polymerization onto polydimethylsiloxane. This approach produced an exceptional hydrophobic surface with a 175° water contact angle. Instead of silicones and fluorocarbons, paraffinic hydrocarbons can also be used to achieve hydrophobicity [48, 49]. Low-density polyethylene (LDPE) is a simple and low-cost way of creating a superhydrophobic coating. Controlling the crystallization duration and nucleation rate resulted in a highly porous superhydrophobic PE surface. However, these hydrophobic coating methods have a problem with coat longevity, which is not satisfactory and is proven to be quite insufficient in the case of cotton material. Another disadvantage is the dangerous impact of fluorine compounds on living species, which can also produce skin irritation [50].

8.4.2 Self-Cleaning Textile Finishing by Hydrophilic/Hydrophobic Nanomaterial

The application of nanotechnology is a relatively recent way of creating self-cleaning effects in textiles [1–3]. Because nanoparticles alone may diffuse on numerous substrates more equally and build ordered morphology [5, 13, 18]. The attachment of monodisperses or aggregates of photocatalytic nanoparticles might improve self-cleaning properties. Similarly, combining micro-nanoparticles with low surface energy materials can improve surface roughness and water repellence, allowing the surface to self-clean [26, 36]. Two types of self-cleaning surfaces can be developed by nanotechnology [51]. The first and foremost, incredibly water-repellent, microscopically rough surfaces. Dirt particles cannot adhere to them and are thus removed. The second example is photocatalytic layers, which use a coating of nanocrystalline titanium dioxide to destroy organic material (such as dirt, contamination, impurities, and microorganisms) when exposed to sunlight [6, 8, 11, 17, 28].

When exposed to sunlight, hydrophilic finishing chemically breaks down dirt and other contaminations, while hydrophobic/superhydrophobic surfaces depend exclusively on the flow of water to wipe the surface [29]. This is known as "photocatalysis." The method is based on the photosynthesis procedure of green leaves, which utilize sunlight to power the chemical. Basically, self-cleaning textile is produced with the help of nanotechnology by photocatalyst, microwaves, carbon nanotubes, metal oxide colloidal, silver nanoparticles, and chlorine halamine [28, 51, 52].

8.4.2.1 Photocatalyst

The photocatalytic process accelerates photoreaction in the existence of a catalyst. The sunlight will be used to break down the dirt molecules in this process. Organic pollutants will be decomposed into air and water by employing the photoreaction caused by photocatalysts [28, 29]. Photoreaction starts when a photocatalyst is exposed to direct light, often ultraviolet light [51, 52]. Electrons of the photocatalyst surface are erected and reached from the valence band to the conduction band. It results in the creation of electron-hole pairs with erected negatively charged electrons (e^-) in the conduction band. There will be positively charged holes (H^+) in the valence band [11, 17, 18]. The formed pairs can gather or become stuck and respond with other materials absorbed by the photocatalyst. The pairs will produce redox reactions at the surface. With negatively charged electrons (e^-) combined with oxygen forms superoxide radical anions (O_2). Positive electric holes combined with water make hydroxyl radicals (OH). The produced highly energetic oxygen species will eventually oxidize organic mixtures to CO_2 and H_2O. As a result, photocatalysts can degrade typical organic materials in the air [6, 8, 52, 53]. The photocatalytic semiconductor technique has shown excellent attributes such as affordable, environmental, and sustainable processing methods. These semiconductors are also used to functionalize various textiles for self-cleaning capabilities. Colored chemicals can be oxidized by functionalized materials [54]. Titanium dioxide (TiO_2) is the most active catalyst in the photodegradation of color stains and other contaminants. It is frequently utilized because of a number of benefits, including safe, accessibility, cost-effectiveness, chemical robustness, and favorable physical qualities [33, 35, 36]. Nanoparticles of TiO_2 and ZnO are utilized to provide self-cleaning and antibacterial characteristics [6–8]. Titanium dioxide is exposed to light with an energy greater than its band gap, and electrons in TiO_2 jump from the valence band to the conduction band. It forms pair of electrons (e^-) and an electric hole (h^+). Negative electrons and oxygen combine to make O_2 radical ions. While positive electric holes and water form hydroxyl radicals (–OH). Because these products are not stable, when an organic compound, such as dirt, contamination, or microorganisms, deposits on the photocatalyst's surface, it will mix with O_2^- and OH^- to form carbon dioxide (CO_2) and water (H_2O). The titanium dioxide acts as a catalyst, continuously breaking down stains repeatedly [54, 55].

The antibacterial properties of TiO_2-enriched cotton fabrics are related to the O_2 radical ions formed during photocatalytic reactions destroying the bacteria cell wall and membrane. Another study found that titanium dioxide-deposited cotton fabrics can remove stains by eliminating the chromophore(s) of the stains when exposed to ultraviolet rays from the sun [22]. The self-cleaning attribute is achieved on cotton fabrics in other experiments by covering them with TiO_2 film and loading them with AgI particles. They allow the fabric to clean using light [56]. The photocatalytic impact of AgI–nanoTiO_2 coated cotton fabric exposed to visible light has much better self-cleaning properties than titanium dioxide-treated cotton fabric. This demonstrates that AgI significantly aids TiO_2's photocatalytic reaction, and the impact lasts for multiple photodecomposition cycles. TiO_2 doping is an excellent way to boost photocatalytic activity, and it's simple to include in CVD or sol-gel procedures [11, 22, 51, 56–58]. Zinc oxide appears to offer an alternative to TiO_2 as a semiconductor material, another photocatalyst, with a similar photocatalysis mechanism to titanium dioxide [59].

ZnO has been intensively explored as a degradation material for environmental contaminants. ZnO has been intensively investigated as a degradation material for ecological contaminants [60]. In the photochemical decomposition of reactive dyes, zinc oxide semiconductor outperforms anatase TiO_2 in photocatalytic reaction efficiency [59]. Powdered zinc oxide is widely utilized for the photocatalytic destruction of colors present in effluent [61]. However, after degradation, there is a recovery difficulty with ZnO powder since some zinc oxide powder is flushed out with the solutions. These ZnO functionalized materials are suitable for both effluent treatment and self-cleaning. Zinc oxide powder can be mounted to a textile substrate. Researchers have done positive work in this regard [59–61]. Photocatalytic color removal and self-cleaning of modified polyester fabric were investigated [62]. In which zinc oxide nanorods were hydrothermally attached to polyester fabric. Due to the photocatalytic action of ZnO, they discovered that the functionalized cloth destroyed the stains and decolored the solution. The fabric withstood repeated solution discoloration. However, its stain removal efficiency decreased with each cycle. The modified polyester fabric offers good rubbing durability but poor washing fastness.

The TiO_2–SiO_2-coated cotton fabric showed a significant photoreaction due to amorphous silica [63]. Chemical spacers can be used to attach TiO_2 to cotton textiles. The TiO_2-loaded cotton textiles demonstrated persistent self-cleaning capabilities. It enables the partial elimination of red wine stains during daylight irradiation, resulting in long-term stable performance [14, 54,

61]. Furthermore, Karimi et al. [64] investigated that level of self-cleaning is greater in the case of cured fabrics using the cross-linking process as compared to cured fabrics using the no crosslink approach. The increased self-cleaning is attributed to a higher percentage and even dispersion of Ti particles on the surface of fabrics made from the cross-linking process. In another study, it was found that electrospun photocatalytic fibers are compatible with non-electrospun photocatalytic fibers [28]. Such composite utilizes the high surface area per unit volume of the material made by coaxial electrospinning. In another study, it was discovered that the zinc oxide and titanium dioxide nanoparticles finishing on cotton textile adds ultra-violate resistance, microbial resistance, soil repellence, and self-cleaning properties [61, 65, 66]. It is also investigated that the self-cleaning ability of cotton fabric is higher with high concentrated nano-TiO_2. The self-cleaning ability and hydrophilicity of wool fabrics can also be successfully enhanced by integrating the TiO_2/SiO_2 nanocomposite. A higher quantity of silica in the TiO_2/SiO_2 nanocomposite results in more efficacy in disintegrating the stains [45]. These functionalized fabrics have better hydrophobicity, anti-soil, anti-bacterial, antistatic, and ultraviolet protection.

8.4.2.2 Microwaves

Because of their weak washing fastness, nanoparticles have a low affinity for textile surfaces [65, 66]. The microwave technique is being developed to attach nanoparticles to the surface of textiles. By connecting various compounds that can repel water, oil, or bacteria, the ability of the nanoparticles is improved toward self-cleaning [67]. These dual nanoparticle-functional compounds form a coating over the textile surface, killing bacteria while repelling liquid and debris. This technology was created by scientists in the United States Air Force and is now being utilized to produce antimicrobial clothing. Such clothes may sustain hygiene for several days without washing [68]. Viet Ha et al. [69] developed a microwave-modified self-cleaning and UV-protecting cotton fabric with a photoactive ZnO layer. They reported that such a ZnO coating layer enhances its hydrophilicity and ultimately leads to self-clean. This functionality was evaluated based on the coffee stain removal efficiency under UV light at various humidity levels (30–90% R_H). Coffee stains on coated materials were almost entirely eliminated after 15 hours without the use of any water or detergent. The highest elimination impact was recorded at 90% R_H.

8.4.2.3 Carbon nanotubes

Carbon nanotubes exhibit numerous amazing qualities, including excellent adsorption, high mouldability, low mass density, high strength, high elasticity, good electromagnetic shielding, and thermal conductivity [45, 52, 62]. The controlled assembly of carbon nanotubes on a textile surface can fulfill the multifunctional requirements. Using this approach, a highly hydrophobic contact angle larger than 150° can be attained. Due to the presence of carbon nanotubes, the electro-conduction of the fabrics increases and may perform several sensory functions [70, 71]. The fabric's self-cleaning function does not weaken after repeated usage. Another investigation found that fluorinated carbon nanotubes outperform regular carbon nanotubes in terms of performance. Vertically organized fluorinated carbon nanotubes on a textile substrate show improved hydrophobicity and outstanding self-cleaning ability [52, 72]. The addition of multi-wall carbon nanotubes to nano-TiO_2 considerably impacted the self-cleaning attributes of the treated cotton fabrics, particularly when exposed to sunshine. Cotton fabric coated with TiO_2/MWCNTs has a high ability to resist UV rays [70–73].

8.4.2.4 Silver nanoparticles

Silver (Ag) nanoparticles are also placed onto the textile substrate for multiple uses. Silver is a powerful antibacterial agent. Its nanoparticles are a highly active substance due to their increased specific area [1]. A fine layer of silver nanoparticles on the fabric results in self-cleaning. Such particles degrade organic molecules such as contamination, pollutants, and microorganisms, resulting in less laundering of clothes. The silver nanoparticles repel water by forming nanowhiskers on the textile surface [7, 74]. The Ag nanoparticles cause a rough surface without affecting strength. Ag nanoparticles are also utilized in combination with TiO_2 particles to treat the surface of textiles. This attachment may be colloidally or specifically to improve functioning. High-temperature curing can promote particle attachment on textile surfaces [75]. Curing at high temperatures on cotton or polyester fibers results in an active surface produced by oxygen that contains a variety of polar groups. Such groups enhance the interaction between Ag and TiO_2 on the textile material. High-frequency plasma with oxygen and vacuum UV are being investigated to improve the adherence of Ag and TiO_2 to textile surfaces [76]. Furthermore, Ag/ZnO coating on the cotton surface also shows good self-cleaning attributes [77, 78].

8.4.2.5 Colloidal metal oxide

Metal oxide–polymer nanocomposites have gotten a lot of attention because of their potential for designing self-cleaning surfaces. For developing such surfaces, a colloidal solution of appropriate metal-oxide particles is to be made. After this textile substrate is immersed in this solution before undergoing heating to develop nanometer-scale roughness on the surface [79]. With this treatment, the cloth becomes hydrophobic with a water contact angle greater than 150°. This colloid suspension approach can coat synthetic textiles with TiO_2. With time, these TiO_2-coated fabrics can eliminate visible light stains such as tea, coffee, and wine [80, 81]. The longevity of the titanium dioxide finish obtained by this approach is adequate for many applications. In addition, a colloidal solution of TiO_2 powder and titanium isopropoxide (TTIP) on wool/polyamide and polyester textiles show more significant results. It gives stable and effective stain removal ability. It can even work in the presence of a neon light. TiO_2–SiO_2 sol-gel can provide a transparent photoactive layer on textiles at low temperatures without affecting the textile surface. It gives a superior photocatalytic impact as compared to TiO_2 alone [81–84].

8.4.2.6 N-halamine

Chlorine is a well-known disinfectant that is utilized to kill microorganisms. The chlorine atom found in N-halamines has been investigated for biocidal purposes [2, 87]. In most cases, the halogen in stable N-halamines is chlorine (N–Cl). The stability of the N–Cl bond is determined by the chlorination reaction that forms the bond. In most cases, amine, amide, and imide groups are chlorinated in a dilute hypochlorite solution. N-halamines are biocides that may destroy a wide variety of microorganisms such as bacteria, fungi, and viruses. This antimicrobial activity is ascribed to the ability of chloride ion (–Cl) in N–Cl to electrophilic substitute for hydrogen ion (–H). In the presence of water (H_2O), this substitution reaction results in the transfer of Cl^+ ions, which can attach to acceptor areas on microorganisms [88]. As a result, microorganisms' enzymatic and metabolic functions are hampered and eliminated. When N-halamines kill bacteria, the N–Cl bonds are changed to N–H bonds, which are inactive and lack antibacterial capabilities. As a result, regeneration is required through treatment with a dilute hypochlorite solution. N-halamines can be used to manufacture antimicrobial textile substrates such as cellulose, polyamide, and polyester. N-halamines include chlorine. However, they are not harmful since toxic chlorine gas is not produced during the process. N-halamine-treated textiles can quickly destroy bacteria on contact, making them ideal for hygiene and medical applications such as uniforms, beds, towels, and wipes [87–89].

8.5 Self-Cleaning Finishes on Different Textile Substrates

The need for functional textile products is increasing rapidly, and it is paving the way for functional finishing. Such functional finishes can lead to abrasion resistance, self-cleaning, antibacterial, soil resistance, and UV protection capabilities for the textile [7, 10, 11, 27]. Self-cleaning fibrous materials are expected to have a great economic possibility in the global scenario. As a result, this new concept continues to provide the intriguing potential for additional research and development [23, 30].

8.5.1 Self-Cleaning Cotton Fibers/Fabrics

Cotton is the largest used fiber in textile products. Recently, the development of superhydrophobic or superhydrophilic coatings on cotton fabric has shown great interest because of its self-cleaning functionalized properties [4]. Under UV irradiation, the anatase nanocrystallite-treated fabrics revealed excellent self-cleaning activity. It also shows antibacterial, colorant disintegration, and disintegration of stains (red wine and coffee) properties [14, 43, 45, 50, 54]. The titania coating absorbed a significant amount of UV light, providing good UV protection to the cotton. A comparative investigation of the ripping strength revealed that there is no impact on the cellulose chains of cotton-coated due to continuous exposure to sunlight [33, 75]. The self-cleaning of red wine stains was used to examine the photocatalytic activity of TiO_2–SiO_2-coated cotton textiles. It was demonstrated that a TiO_2–SiO_2 species with appropriate photoactivity on non-heat resistant materials could be formed at temperatures of 100°C. In the study, the highest titanium dioxide and SiO_2 concentration is 5.8 and 3.9 percent (w/w), respectively. The CO_2 evolution due to stain degradation is high in the case of TiO_2–SiO_2 treated samples than in TiO_2-treated cotton samples [90]. No impact on the textile surface is also found even after several degradation cycles. The thickness of TiO_2–SiO_2 coating is found 20 to 30 nanometers. The particle sizes of TiO_2 and SiO_2 were found from 4 to 8 nanometers. Further SEM and energy-dispersive spectroscopy (EDS) investigation revealed that the Ti particles did not remain alone. They were surrounded by amorphous SiO_2 [55, 82, 91].

8.5.2 Self-Cleaning of Protein Fibers/Fabrics

Wool and silk are natural protein-based fabrics that are highly sought after, rich, delicate, and commonly utilized in the clothing industry. However, the cleaning process of such fiber is delicate. These fibers are more prone to attack by detergents and soaps. Silk and wool have poor thermal and chemical stability. They also show poor resistance to exposure to UV-light and are more susceptible to microorganism attacks than cellulose fibers [51, 55]. As a result, protein fibers must have self-cleaning properties [55, 58, 80]. Nanotechnology has enabled the development of self-cleaning wool and silk. In the sunshine, wool and silk products would soon clean themselves from odors and contaminations [1, 7, 18]. The key is a nanoparticle coating, which is already used for several other textile products [18, 19].

Wool fibers treated with anatase nanocrystals of TiO_2 are used through the sol-gel process. It provides self-cleaning and UV protection to the material [29, 51, 55]. Colloids of anatase TiO_2 can also be formed at room temperature [57]. This breakthrough is helpful for low thermal resistance materials [1, 27, 30, 40]. It was also discovered hydrochloric acid could develop an aqueous colloid of TiO_2 in the form of a single-phase anatase of 4 to 5 nm size. These nanosized crystals can be applied to a keratin fiber sample with no change in its intrinsic qualities [29, 92].

TiO_2 has a strong affinity for hydroxyl and carboxylic groups, particularly the latter. On the other hand, keratin fibers have fewer than half of those functional groups. As a result, these fibers may not show stable bonds with the anatase nanocrystals [29, 92, 93]. Acylation of fibers is one approach for carboxylic group enrichment. Thus, acylation by acid enhances the binding ability toward metal ions [81]. Succinic anhydride is a nontoxic and mild acylating agent. Acylation further enhances self-cleaning functionality. The acylation process has paved the new application area for protein fibers [29, 33]. Lots of studies were done by several researchers in this respect [29, 33, 53, 56].

8.5.3 Self-Cleaning Polyester Fibers/Fabrics

Polyesters fiber is a highly used synthetic fiber and has found widespread application. Functionalizing polyester fiber has enriched its worthiness [13]. It is also used in self-cleaning activities. Adhesion of TiO_2 particles to frequently washed cloths leads to fewer washing requirements. But the bonding between TiO_2 and polyester is poor due to a lack of chemical affinity. There is a requirement for surface treatment to change the affinity. The most adaptable strategy for improving the bondability of TiO_2 on polyester fibers is probably low-temperature plasma (LTP) pretreatment. Surface pretreatment only affects the polymer's outermost surface layers and does not affect its bulk properties [33, 36, 58]. A comparative study shows that direct current glow discharge plasma gives the best results for functionality enhancement as compared to plasma and microwave plasma [58, 62]. The most commonly used gas in plasma treatment is oxygen gas, which introduces negative groups COO–, –O–O– onto the polyester surface. It results in a highly oxygenated surface and enhanced bondability [12–14, 30]. Such modified polyester provides significant improvement in self-cleaning activity. The titania-coated polyester's UV absorption was significant enough to support outstanding UV protection for polyester [33, 36, 58, 61, 62].

8.5.4 Self-Cleaning Modified Cellulose Fibers/Fabrics

In several investigations, coating TiO_2 nanoparticles and composites on the modified cellulose products have shown promising results in regard to the self-cleaning activity. Regenerated cellulose fabrics provide a soft handle and a good appearance if they are treated with TiO_2 nanoparticles and TiO_2/SiO_2 composite [80, 94]. M. A. Ramadan [95] applied TiO_2-nanosol coating to cellulose acetate fabrics pretreated with H_2O_2. The self-cleaning ability of the fabrics is enhanced by increasing the concentration of TiO_2-nanosol to a particular level, extending the curing period for 15 sec., and increasing the microwave power from 80% to 100%, with a confidence of 90%. Furthermore, UV radiation for up to 90 minutes is required to achieve outstanding self-cleaning capabilities while retaining other physical and mechanical attributes. If the binders are used in the finishing pad bath to stabilize titanium dioxide deposition, they show superior self-cleaning. Pretreatment of cellulose acetate textiles with H_2O_2 is required to ensure durability.

8.6 Evaluation of Self-Cleaning Textiles

Surface morphology is an important aspect of characterization the of self-cleaning fabrics. Researchers have developed a number of evaluation methods to investigate surface morphology [1, 7, 23, 59]. This section covers various ways to analyze the materials, surface chemistry, and self-cleaning activities.

The contact angle is an important criterion for evaluating the wettability of self-cleaning textiles, which reflects the hydrophobicity and hydrophilicity of the surface. Apart from the contact angle measurement, surface tension measurement is another important factor. The surface roughness of self-cleaning textiles is measured by atomic force microscopy (AFM) topography. Environmental ellipsometric porosimetry, grazing incidence X-ray analyses at low and wide angles (GI-SAXS and GI-WAXS), electronic and near-field microcopies, field-emission scanning electric microscopy (FE-SEM), transmission electron microscopy (TEM), Fourier-transform infrared spectroscopy (FTIR), UV-visible transmittance, X-ray photoelectron spectroscopy (XPS) and differential scanning calorimetry (DSC) are a few other techniques that are widely used in characterizing self-cleaning textiles [54, 59, 96, 97].

Photocatalytic action is also another important function for self-cleaning fabrics. The photocatalytic efficacy of self-cleaning fabrics is typically evaluated by degrading organic contaminations. This method is known as the pollutant degradation method. On the other hand, the photodegradation of the colorant is evaluated based on decomposition activities, such as solution discoloration and stain degradation. In the solution discoloration, self-cleaning fabrics are dipped into the dye solution and irradiated to UV light. The concentration of this dye solution is periodically measured by the UV-Vis spectrophotometer, and efficiency is calculated [62, 96].

In stain degradation evaluation, the fabric is stained and exposed to ultraviolet light for a specific period. The sample is examined for color strength (K/S) values by the color spectrometer. Quantitative assessment of self-cleaning action can be measured by comparison of color strength values of the exposed and normal portions of the same stain. The decline in color strength values shows that the stains are removed [54, 96–98]. Besides color strength value, such a fabric can be evaluated by gas chromatography [83, 84, 86].

The UV resistivity of self-cleaning fabrics is determined by the ultraviolet protection factor values (UPF). Generally, UPF tests are done by the spectrophotometer or a spectroradiometer [2, 22, 100]. Sivakumar et al. [101] assessed UPF for self-cleaning fabric made by nano-ZnO and nano-TiO_2 with an acrylic binder. They found that the UPF of the fabrics comprising larger TiO_2 nanoparticles is higher than smaller-sized ZnO nanoparticles.

8.7 Applications of Self-Cleaning Textiles

Self-cleaning textiles have a wide range of applications, starting from clothing to automobile, optical, marine industry, and aerospace industries. After the chemical treatment, self-cleaning products retain their original aesthetic and handle values. There is no need to frequently wash for the self-cleaning textile. With the help of this, they save a lot of energy, water, and soap. The technologies discussed in the chapter can prevent dirt, contamination, soil, and oil deposition on the self-cleaning textile [1, 2, 19, 22, 30, 101]. Anti-dirt, antibacterial and self-cleaning wear can widely be used in various technical textiles areas. Self-cleaning finishing can also be utilized in outdoor curtains, covers, exterior decoration, and paintings. Diversified applications of self-cleaning surfaces, numerous manufacturers are utilizing this technology and commercializing it in the form of products [18, http://www. pilkington.com/; http://www.ppg.com/en/Pages/default.aspx; http://www.lotusan.de/]. Few commercial goods with this self-cleaning quality are available on the global market. Mincor® TX TT, for example, can be utilized for designing outdoor textiles such as shading nets, ribbons, rain protectors, canopies, and rain shelters. NanoTex® is utilized for clothing such as kurtas, shirts, pinafores, coats, muffs, suiting, and so forth (https://www.basf.com/ru/ru/media/news-releases/2009/06/p-09-263.html). Nanosphere® is primarily utilized in the preparation of self-cleaning men's shirts [https://www.bluesmiths.com/pages/fabrics-and-technologies].

The textile materials thus resist water droplets, and dirt particles can be washed off automatically. When fabrics require no or very little washing, this result is self-cleaning. Even washing conditions will be soft at low temperatures, with only a small amount of soap or detergent. These items are also extremely durable and long-lasting. Even after multiple washing cycles, the texture, hand feel, and comfortability remain unaffected. As a result, NanoSphere® is positioned as an excellent product for use in outdoor garments, sportswear, men's and women's wear, workwear, shoe covers, and home furnishings [https://www.schoeller-textiles.com/en/technologies/nanosphere].

8.8 Limitations of Self-Cleaning Fabric

Several research projects are effective in the laboratory, yet they miss the mark catastrophically when translated into bulk or actual products for a variety of reasons. Despite major achievements in manufacturing self-cleaning textile fabric, there are difficulties and challenges. According to Li et al. [39], the primary mechanisms of the textile and chemical reagents throughout the fabrication process are challenging to handle. Most fabrication processes are only applicable in the laboratory and are not appropriate for large-scale industrial output. Water-repellent fabric's mechanical durability suffers from weak mechanical stability and durability during laundry. Although numerous studies have been conducted successfully in the laboratory, many have failed when applied to real-time goods. The "peeling off" of the coatings from the surface is a foremost problem. This should always be in mind during the development phase of the product [19, 27, 39].

To avoid air gaps, the coating should be placed over the surface. The resulting coating should not enable water droplets to collect on the surface but rather should roll off. The coating's surface should be free of micro fractures and scratches. The layer should be put onto the substrate with no air gaps [19, 27, 44]. Another aspect that must be addressed is the behavior of water droplets on the surface of the coating. The coating should not enable water droplets to pass through it [7, 44]. These are some of the other factors that will necessitate additional study and development in order to make self-cleaning coatings commercially viable. Self-cleaning textiles are becoming more feasible and cost-effective as nanoscience and nanotechnology improve [7, 27, 30].

In addition to the limitations, the washing fastness self-cleaning finishing material is a major concern [27, 56, 81, 82, 104, 105]. Not all of the agents are appropriate for all textile substrates. Another impact is the amount of time spent and the intensity of direct sunlight. For example, there is a requirement for a whole day of sunlight to remove a tea stain [23]. Similarly, electron excitation in the valence band of TiO_2 is dependent on the intensity of sunlight. If the intensity is not proper, the cleaning process does not start. The electrons in the inversion band must react with ambient oxygen, causing oxygen depletion and raising environmental concerns [27].

8.9 Conclusion

The manufacturing of fabrics with a self-cleaning surface has been intensively explored, and the development of unique functional textile materials has piqued researchers' interest. Surfaces with reversible wettability are critical for a wide range of industrial applications. The development of self-cleaning surfaces in this context is based on the lotus effect (hydrophobic) or photocatalysts (hydrophilic). Nanotechnology offers a new way of self-cleaning textiles that provide fresh clothes every day, which is economically beneficial. Silver, TiO_2, ZnO, and other nanosized metal and metal-oxide particles have been synthesized and coated to various textile substrates to develop self-cleaning and antibacterial characteristics. In addition to self-cleaning features, such materials have intrinsic hydrophobicity, UV resistance, antistatic, oil repellency, and abrasion resistance. The multifunctional coating is an open area for further investigation. Apart from clothing, self-cleaning surfaces can be used for various purposes, including bio-fouling in medical instruments, building and car windows, and solar panel cover glass in the outside environment.

References

1. Shah, M. A., Pirzada, B. M., Price, G., Shibiru, A. L., and Qurashi, A. (2022). Applications of nanotechnology in smart textile industry: A critical review. *Journal of Advanced Research*, 38, 55–75.
2. Khan, I., Saeed, K., and Khan, I. (2019). Nanoparticles: Properties, applications, and toxicities. *Arabian Journal of Chemistry*, 12, 908–931.
3. Boyce, J. M. (2016). Modern technologies for improving cleaning and disinfection of environmental surfaces in hospitals. *Antimicrobial Resistance and Infection Control*, 5, 10. https://doi.org/10.1186/s13756-016-0111-x
4. Sun, D. and Böhringer, K. (2019). Self-cleaning: From bio-inspired surface modification to MEMS/microfluidics system integration. *Micromachines*, 10(2), 101.
5. Somasundaram, S. and Kumaravel, V. (2019). Application of nanoparticles for self-cleaning surfaces. In: Rajendran, S., Naushad, M., Raju, K., and Boukherroub, R. (eds.) *Emerging Nanostructured Materials for Energy and Environmental Science. Environmental Chemistry for a Sustainable World*, Vol. 23, Springer, Cham. https://doi.org/10.1007/978-3-030-04474-9_11
6. Ganesh, V. A., et al. (2012). Photocatalytic superhydrophilic TiO_2 coating on glass by electrospinning. *RSC Advances*, 2(5), 2067. https://doi.org/10.1039/c2ra00921h
7. Maity, S., Singha, K., and Pandit, P. (2020). Self-cleaning finishes for functional and value added textile materials. In: Shahid, M. and Adivarekar, R. (eds.), *Advances in Functional Finishing of Textiles. Textile*

Science and Clothing Technology. Springer, Singapore, pp. 217–229. https://doi.org/10.1007/978-981-15-3669-4_9

8. Banerjee, S., Dionysiou, D. D., and Pillai, S. C. (2015) Self-cleaning applications of TiO_2 by photoinduced hydrophilicity and photocatalysis, *Applied Catalysis B: Environmental*, 176, 396–428.
9. Johnson, R. E. and Dettre, R. H. (1969). Wettability and contact angles. In: Matijevic, E. (ed.), *Surfaces and Colloids*, Vol. 2, Wiley-Interscience, New York, p. 85.
10. Shafrin, E. G. and Zisman, W. A. (1960). Constitutive relations in the wetting of low energy surfaces and the theory of the retraction method of preparing monolayers. *Journal of Physical Chemistry*, 64, 519–524.
11. Foster, H. A., Ditta, I. B., Varghese, S., and Steele, A. (2011). Photocatalytic disinfection using titanium dioxide: Spectrum and mechanism of antimicrobial activity. *Applied Microbiology and Biotechnology*, 90, 1847–1868. https://doi.org/10.1007/s00253-011-3213-7
12. Haji, A. and Kan, C.-W. (2021). Plasma treatment for sustainable functionalization of textiles. In: Ibrahim, N. and Hussain, C. M. (eds.), *The Textile Institute Book Series, Green Chemistry for Sustainable Textiles*, Woodhead Publishing, pp. 265–277.
13. Mihailovic, D., Saponjic, Z., Molina, R., et al. (2010) Improved properties of oxygen and argon R.F. plasma-activated polyester fabrics loaded with TiO_2 nanoparticles. *ACS Applied Materials Interfaces*, 2, 1700–1706.
14. Mejia, M. I., Marin. J. M., Restrepo, G., et al. (2009). Self-cleaning modified TiO_2-cotton pretreated by UVC-light (185 nm) and RF-plasma in vacuum and also under atmospheric pressure. *Applied Catalysis B: Environmental*, 91, 481–488.
15. Gulrajani, M. L. (2006). Nano finishes. *Indian Journal of Fibre and Textile Research*, 31, 187–201.
16. Nakajima, A., Hashimoto, K., Watanabe, T., Takai, K., Yamauchi, G., and Fujishima, A. (2000). Transparent superhydrophobic thin films with self-cleaning properties. *Langmuir*, 16, 7044–7047.
17. Paz, Y., Luo, Z., Rabenberg, L., and Heller, A. (1995). Photooxidative self-cleaning transparent titanium dioxide films on glass. *Journal of Materials Research*, 10(11), 2842–848.
18. Shen, W., Zhang, C., Li, Q., Zhang, W., Cao, L., and Ye, J. (2015). Preparation of titanium dioxide nanoparticle modified photo-catalytic self-cleaning concrete. *Journal of Cleaner Production*, 87, 762–765.
19. Li, T., Ren, T., and He, J. (2016). The inspiration of nature: Natural counterparts with self-cleaning functions. In: He, J. (ed.), *Self-cleaning*

Coatings: Structure, Fabrication, and Application, pp. 1–24. https://doi.org/10.1039/9781782623991

20. Barthlott, W. and Neinhuis, C. (1997). Purity of the sacred lotus or escape from contamination in biological surfaces. *Planta*, 202, 1–8. https://doi.org/10.1007/s004250050096
21. Koch, K., Bhushan, B., and Barthlott, W. (2009). Multifunctional surface structures of plants: An inspiration for biomimetics. *Progress in Materials Science*, 54(2), 137–178.
22. Ibrahim, A., Laquerre, J., Forcier, P., Deregnaucourt, V., Decaens, J., and Vermeersch, O. (2021). Antimicrobial agents for textiles: Types, mechanisms, and analysis standards. In: Kumar, B. (ed.), *Textiles for Functional Applications*, IntechOpen. https://doi.org/10.5772/intechopen.98397
23. Atwah, A. A. and Khan, M. A. (2022). Influence of microscopic features on the self-cleaning ability of textile fabrics. *Textile Research Journal*, 93(1–2), https://doi.org/10.1177/00405175211069881
24. Owais, A., Khaled, M., Yilbas, B. S. (2017). Hydrophobicity and surface finish. In: Hashmi, M. S. J. (ed.), *Comprehensive Materials Finishing*, Elsevier, pp. 137–148, ISBN 9780128032497, https://doi.org/10.1016/B978-0-12-803581-8.09172-4
25. Chen, W., et al. (1999). Ultrahydrophobic and ultralyophobic surfaces: Some comments and examples, *Langmuir*, 15, 3395–3399.
26. Rios, P. F., Dodiuk, H., Kenig, S., McCarthy, S., and Dotan, A. (2006). The effects of nanostructure and composition on the hydrophobic properties of solid surfaces. *Journal of Adhesion Science and Technology*, 20(6), 563–587. Doi: 10.1163/156856106777213302
27. Dalawai, S. P., et al. (2020). Recent advances in durability of superhydrophobic self-cleaning technology: a critical review. *Progress in Organic Coating*, 138, 1–13.
28. Bedford, N. M. and Stecki, A. J. (2010) Photocatalytic self-cleaning textile fibers by coaxial electrospinning. *ACS Applied Materials Interfaces*, 2(8), 2448–2455.
29. Tung, W. S. and Daoud, W. A. (2009). Photocatalytic self-cleaning keratins: A feasibility study. *Acta Biomaterialia*, 5, 50–56.
30. Jelil, R. A. (2015). A review of low-temperature plasma treatment of textile materials. *Journal of Materials Science,* 50, 5913–5943. https://doi.org/10.1007/s10853-015-9152-4
31. Hegemann, D. (2006). Plasma polymerization and its applications in textiles. *Indian Journal of Fibre and Textile Research*, 31, 99–115.

32. Morent, R., De Geyter, N., and Verschuren, J. et al (2008). Non-thermal plasma treatment of textiles. *Surface and Coatings Technology*, 202, 3427–3449.

33. Qi, K., Xin, J. H. and Daoud, W. A., et al. (2007). Functionalizing polyester fiber with a self-cleaning property using anatase TiO_2 and low-temperature plasma treatment. *International Journal of Applied Ceramic Technology*, 4, 554–563.

34. Kale, K. H. and Desai, A. N. (2011). Atmospheric pressure plasma treatment of textiles using non-polymerising gases. *Indian Journal of Fibre and Textile Research*, 36, 289–299.

35. Baghriche, O., Rtimi, S., and Pulgarin, C., et al. (2013). RF-plasma pretreatment of surfaces leading to TiO_2 coatings with improved optical absorption and OH-radical production. *Applied Catalysis B: Environmental*, 130–131 65–72.

36. Hashemizad, S., Haji, A., and Mireshghi, S. S. (2014). Environmentally friendly plasma pretreatment for preparation of self-cleaning polyester fabric with enhanced deposition of TiO_2 nanoparticles. *Journal of Biodiversity and Environmental Science*, 5, 220–226.

37. Rivero, P. J. , Vicente, A., and Rodriguez, R. J. (2020). Electrospinning technique as a powerful tool for the design of superhydrophobic surfaces. In: Pham, P., Goel, P., Kumar, S., and Yadav, K. (eds.), *21st Century Surface Science: A Handbook*. IntechOpen. https://doi.org/10.5772/intechopen.92688

38. Acik, G., Cansoy, C. E., and Kamaci, M. (2019). Effect of flow rate on wetting and optical properties of electrospun poly(vinyl acetate) micro-fibers. *Colloid and Polymer Science*, 297(1), 77–83.

39. Li, D. and Xia, Y. (2004). Electrospinning of nanofibers: Reinventing the wheel? *Advanced Materials*,16(14), 1151–1170.

40. Kurusu, R. S. and Demarquette, N. R. (2019). Surface modification to control the water wettability of electrospun mats. *International Materials Review*, 64(5), 249–287.

41. Su, D., Huang, C., Hu, Y., Jiang, Q., Zhang, L., and Zhu, Y. (2011). Preparation of superhydrophobic surface with a novel sol-gel system. *Applied Surface Science*, 258(2), 928–934, ISSN 0169-4332. https://doi.org/10.1016/j.apsusc.2011.09.030

42. Zhang, Y., Dong, B., Wang, S., Zhao, L., Wan, L., and Wang, E. (2017). Mechanically robust, thermally stable, highly transparent superhydrophobic coating with low-temperature sol-gel process. *RSC Advances*, 7(75), 47357–47365. Doi: 10.1039/c7ra08578h

43. Xue, C.-H., Jia, S.-T., Chen, H.-Z., Wang, M. (2008). Superhydrophobic cotton fabrics prepared by sol-gel coating of TiO_2 and surface hydrophobization. *Science and Technology of Advanced Materials*, 9(3), 035001. Doi: 10.1088/1468-6996/9/3/035001
44. Liu, J., Huang, W., and Xing, Y., et al. (2011). Preparation of durable superhydrophobic surface by sol-gel method with water glass and citric acid. *Journal of Sol-Gel Science and Technology*, 58, 18–23. https://doi.org/10.1007/s10971-010-2349-8
45. Zhang, L., Xue, C. H., Cao, M., Zhang, M. M., Li, M., and Ma, J. Z. (2017). Highly transparent fluorine-free superhydrophobic silica nanotube coatings. *Chemical Engineering Journal*, 320, 244–252.
46. Liu, S., Liu, X., Latthe, S. S., Gao, L., An, S., Yoon, S. S., Liu, B., and Xing, R. (2015). Self-cleaning transparent superhydrophobic coatings through simple sol-gel processing of fluoroalkyl silane. *Applied Surface Science*, 351, 897–903.
47. Syafiq, A., Vengadaesvaran, B., Rahim, N. A., Pandey, A. K., Bushroa, A. R., Ramesh, K., and Ramesh, S. (2019). Transparent self-cleaning coating of modified polydimethylsiloxane (PDMS) for real outdoor application. *Progress in Organic Coatings*, 131, 232–239. ISSN 0300–9440. https://doi.org/10.1016/j.porgcoat.2019.02.020
48. Liu, J., Yao, Y., Li, X., and Zhang, Z. (2021). Fabrication of advanced polydimethylsiloxane-based functional materials: Bulk modifications and surface functionalizations. *Chemical Engineering Journal*, 408, 127262. https://doi.org/10.1016/j.cej.2020.127262
49. Periyasamy, A. P., Venkataraman, M., Kremenakova, D., Militky, J., and Zhou, Y. (2020). Progress in sol-gel technology for the coatings of fabrics. *Materials*, 13(8), 1838. https://doi.org/10.3390/ma13081838
50. Foorginezhad, S. and Zerafat, M. M. (2019). Fabrication of superhydrophobic coatings with self-cleaning properties on cotton fabric based on octa vinyl polyhedral oligomeric silsesquioxane/polydimethylsiloxane (OV-POSS/PDMS) nanocomposite. *Journal of Colloid and Interface Science*, 540, 78–87, ISSN 0021-9797. https://doi.org/10.1016/j.jcis.2019.01.007
51. American Chemical Society. Nanotechnology lends a hand with 'self-cleaning' wool and silk fabrics. *ScienceDaily*. www.sciencedaily.com/releases/2008/02/080211094258.htm (accessed May 31, 2022)
52. Saleh, T. (2013). The role of carbon nanotubes in enhancement of photocatalysis. In: Suzuki, S. (ed.), *Syntheses and Applications of Carbon Nanotubes and Their Composites*, IntechOpen, London, 10.5772/51050

53. Yousif, E., and Haddad, R. (2013). Photodegradation and photo stabilization of polymers, especially polystyrene: A review. *SpringerPlus*, 2, 398. https://doi.org/10.1186/2193-1801-2-398

54. Chaudhari, S. B., Mandot, A. A., and Patel, B. H. (2012). Effect of nano TiO_2 pretreatment on functional properties of cotton fabric, *IJERD*, 1(9), 24–29.

55. Nitayaphat, W., Jirawongcharoen, P., and Trijaturon, T. (2018). Self-cleaning properties of silk fabrics functionalized with TiO_2/SiO_2 composites. *Journal of Natural Fibers*, 15, 2, 262–272.

56. Kalia, S., Dufresne, A., Cherian, B. M., Kaith, B. S., Avérous, L., Njuguna, J., and Nassiopoulos, E. (2011). Cellulose-based bio- and nanocomposites: A review. *International Journal of Polymer Science*, 2011, 1–35. https://doi.org/10.1155/2011/837875

57. Anucha, C. B., Altin, I., Bacaksiz, E., and Stathopoulos, V. N. (2022). Titanium dioxide (TiO_2)-based photocatalyst materials activity enhancement for contaminants of emerging concern (CECs) degradation: In the light of modification strategies. *Chemical Engineering Journal Advances*, 10, 100262, ISSN 2666–8211. https://doi.org/10.1016/j.ceja.2022.100262

58. Haji, A., Shoushtari, A. M., Mazaheri, F., and Tabatabaeyan. S. E. (2016). RSM optimized self-cleaning nano-finishing on polyester/wool fabric pretreated with oxygen plasma. *Journal of the Textile Institute*, 107, 8, 985–994.

59. Zhu, C., Shi, J., Xu, S., Ishimori, M., Sui, J., and Morikawa, H. (2017). Design and characterization of self-cleaning cotton fabrics exploiting zinc oxide nanoparticle-triggered photo-catalytic degradation. *Cellulose*, 24, 2657–2667. https://doi.org/10.1007/s10570-017-1289-7

60. Soltaninezhad, M. and Aminifar, A. (2011). Study nanostructures of semiconductor zinc oxide (ZnO) as a photocatalyst for the degradation of organic pollutants. *International Journal Nano Dimension*, 2(2), 137–145.

61. Anita, S., Ramachandran, T., Koushik, C. V., Rajendran, R., and Mahalakshmi, M. (2010). Preparation and characterization of zinc oxide nanoparticles and a study of the antimicrobial property of cotton fabric treated with the particles. *Journal of Textile and Apparel Technology Management*, 6(4), 1–3.

62. Ashraf, M., Champagne, P., Perwuelz, A., et al. (2015). Photo-catalytic solution discoloration and self-cleaning by polyester fabric

functionalized with ZnO nanorods. *Journal of Industrial Textiles*, 44, 884–898.

63. Li, Z., Dong, Y., Li, B., Wang, P., Chen, Z., and Bian, L. (2018). Creation of self-cleaning polyester fabric with TiO_2 nanoparticles via a simple exhaustion process: Conditions optimization and stain decomposition pathway. *Materials and Design*, 140, 366–375, ISSN 0264-1275.

64. Karimi, L., Mirjalili, M., Yazdanshenas, M. E., and Nazari, A. (2010). Effect of nano TiO_2 on self-cleaning property of cross-linking cotton fabric with succinic acid under UV irradiation. *Photochemistry and Photobiology*, 86, 1030–1037. https://doi.org/10.1111/j.1751-1097.2010.00756.x

65. Abidi, N., Cabrales, L., and Hequet, E. (2009). Functionalization of a cotton fabric surface with titania nanosols: Applications for self-cleaning and UV-protection properties. *ACS Applied Materials Interfaces*, 1, 2141–2146.

66. Eglītis, R. and Mežinskis, G. (2017). Comparison of treatments of a cotton fabric modified with a low-temperature TiO_2 coating. *Proceedings of the Estonian Academy of Sciences*, 66(4), 473. https://doi.org/10.3176/proc.2017.4.21

67. Kar, E., Bose, N., Dutta, B., Mukherjee, N., Mukherjee, S. (2019). Ultraviolet- and microwave-protecting, self-cleaning e-skin for efficient energy harvesting and tactile mechanosensing. *ACS Applied Materials Interfaces*, 11(19), 17501–17512.

68. Li, Z., Zheng, Q., Wang, Z. L., and Li, Z. (2020). Nanogenerator-based self-powered sensors for wearable and implantable electronics. *Research*, 2020, 8710686.

69. Thi, V. H. T. and Lee, B.-K. (2017). Development of multifunctional self-cleaning and UV blocking cotton fabric with modification of photoactive ZnO coating via microwave method. *Journal of Photochemistry and Photobiology A: Chemistry*, 338, 13–22, ISSN 1010-030. https://doi.org/10.1016/j.jphotochem.2017.01.020

70. Lee, H. J., Kim, J., and Park, C. H. (2013). Fabrication of self-cleaning textiles by TiO_2-carbon nanotube treatment. *Textile Research Journal*, 84, 267–278. Doi: 10.1177/0040517513494258

71. Alamer, F. A., Alnefaie, M. A., and Salam, M. A. (2022). Preparation and characterization of multi-walled carbon nanotubes-filled cotton fabrics. *Results in Physics*, 33, 105205, ISSN 2211-3797. https://doi.org/10.1016/j.rinp.2022.105205

72. Karimi, L., Zohoori, S., and Amini, A. (2014). Multi-wall carbon nanotubes and nano titanium dioxide coated on cotton fabric for superior self-cleaning and UV blocking. *New Carbon Materials*, 29, 380–385, ISSN 1872-5805. https://doi.org/10.1016/S1872-5805(14)60144-X

73. Xu, Q., Zhang, W., Dong, C., Sreeprasad, T. S., and Xia, Z. (2016). Biomimetic self-cleaning surfaces: Synthesis, mechanism, and applications. *Journal of the Royal Society Interface*, 13, 20160300. http://dx.doi.org/10.1098/rsif.2016.0300

74. Shateri-Khalilabad, M., Yazdanshenas, M. E., and Etemadifar, A. (2017). Fabricating multifunctional silver nanoparticles-coated cotton fabric. *Arabian Journal of Chemistry*, 10, S2355–S2362, ISSN 1878-5352. https://doi.org/10.1016/j.arabjc.2013.08.013

75. Tudu, B. K., Sinhamahapatra, A., and Kumar, A. (2020). Surface modification of cotton fabric using TiO_2 nanoparticles for self-cleaning, oil-water separation, anti-stain, anti-water absorption, and antibacterial properties. *ACS Omega*, 5, 7850–7860.

76. Huang, C., Cai, Y., Chen, X., and Ke, Y. (2021). Silver-based nanocomposite for fabricating high performance value-added cotton. *Cellulose*, 29, 723–750. https://doi.org/10.1007/s10570-021-04257-z

77. El-Shishtawy, R. M., Asiri, A. M., Abdelwahed, N. A. M., and Al-Otaibi, M. M. (2011). In situ production of silver nanoparticle on cotton fabric and its antimicrobial evaluation. *Cellulose*, 18, 75–82.

78. Gao, D. G., Liu, J. J., Lyu, L. H., Li, Y. J., Ma, J. Z., and Baig, W. (2020). Construct the multifunction of cotton fabric by synergism between nano ZnO and Ag. *Fibers and Polymers*, 21, 505–512.

79. Krishna, M. G., Vinjanampati, M., and Purkayastha, D. D. (2013). Metal oxide thin films and nanostructures for self-cleaning applications: Current status and future prospects. *The European Physical Journal Applied Physics*, 62(3), 30001. Doi: 10.1051/epjap/2013130048

80. Mariam, D. and Ahmed, G. H. (2021). Self-cleaning properties of cellulosic fabrics (a review). *Biointerface Research in Applied Chemistry*, 12(2), 1847–1855. https://doi.org/10.33263/briac122.18471855

81. Radetić, M. (2013). Functionalization of textile materials with TiO_2 nanoparticles. *Journal of Photochemistry and Photobiology C: Photochemistry Reviews*, 16, 62–76.

82. Sadr, F. A. and Montazer, M. (2014). In situ sonosynthesis of nano TiO_2 on cotton fabric. *Ultrasonics Sonochemistry*, 21, 681–691.

83. Jiang, C., Liu, W., Yang, M., Zhang, F., Shi, H., Liu, W., and Wang, Z. (2019). Robust fabrication of superhydrophobic and photo-catalytic self-cleaning cotton textiles for oil–water separation via thiol-ene click reaction. *Journal of Materials Science*, 54, 7369–7382.

84. Li, S., Huang, J., Ge, M., Cao, C., Deng, S., Zhang, S., Chen, G., Zhang, K., Al-Deyab, S. S., and Lai, Y. (2015). Robust flower-like TiO_2@cotton fabrics with special wettability for effective self-cleaning and versatile oil/water separation. *Advanced Materials Interfaces*, 2, 1–10. https://doi.org/10.1002/admi.201500220

85. Wang, F., Huang, L., Zhang, P., Si, Y., Yu, J., and Ding, B. (2020). Antibacterial N-halamine fibrous materials. *Composites Communications*, 22, 100487. https://doi.org/10.1016/j.coco.2020.100487

86. Turaga, U., Singh, V., and Ramkumar, S. (2015). Biological and chemical protective finishes for textiles. In: Paul, R. (ed.), *Functional Finishes for Textiles*, Woodhead Publishing, pp. 555–578, ISBN 9780857098399, https://doi.org/10.1533/9780857098450.2.555

87. Natan, M., Gutman, O., Lavi, R., Margel, S., and Banin, E. (2015). Killing mechanism of stable N-halamine cross-linked polymethacrylamide nanoparticles that selectively target bacteria, *ACS Nano*, 9, 1175–1188.

88. Cerkez, I., Kocer, H. B., Worley, S. D., Broughton, R. M., and Huang, T. S. (2012). N-halamine biocidal coatings via a layer-by-layer assembly technique, *Langmuir*, 72, 4091–4097.

89. Nazi, N., Marguier, A., Debiemme-Chouvy, C. Humblot, V. (2022). Optimization and antibacterial response of N-halamine coatings based on polydopamine. *Colloids Interfaces*, 6, 9. https://doi.org/ 10.3390/colloids6010009

90. Ramadan, M. A., Raslan, W. M., El-Khatib, E. M., Hebeish, A. (2012). Rendering of cellulose acetate fabrics self-cleaning through treatment with TiO_2 nano particles, *Materials Sciences and Applications*, 3, 872–879. Doi: 10.4236/msa.2012.312127

91. Ahmad, I., Kan, C., and Yao, Z. (2019). Reactive blue-25 dye/TiO_2 coated cotton fabrics with self-cleaning and UV blocking properties. *Cellulose*, 26, 2821–2832. https://doi.org/10.1007/s10570-019-02279-2

92. Tung, W. S. and Daoud, W. A. (2010). Self-cleaning wool: Effect of formulation concentration on low stress mechanical and surface properties. *Research Journal of Textile and Apparel*, 14(2), 83–88.

93. Montazer, M. and Pakdel, E. (2011). Self-cleaning and color reduction in wool fabric by nano titanium dioxide. *Journal of the Textile Institute*, 102, 343–352. Doi: 10.1080/00405001003771242

94. Kwon, S., Ko, T.-J., Yu, E., Kim, J., Moon, M.-W., and Park, C. H. (2014). Nanostructured self-cleaning lyocell fabrics with asymmetric wettability and moisture absorbency (part I), *RSC Advances*, 4, 45442–45448.

95. Ramadan, M. A., Raslan, W. M., El-Khatib, E. M., and Hebeish, A. (2012). Rendering of cellulose acetate fabrics self-cleaning through treatment with TiO_2 nano particles. *Materials Sciences and Applications*, 3, 872–879. https://doi.org/10.4236/msa.2012.312127

96. Jeong, E., Woo, H., Cho, S., and Bae, J.-S. (2018). Preparation and evaluation of self-cleaning fabrics using photocatalyst and superhydrophobic finishing. *Textile Coloration and Finishing*, 30(4), 288–293. https://doi.org/10.5764/TCF.2018.30.4.288

97. Wu, D., Long, M., Zhou, J., Cai, W., Zhu, X., Chen, C., and Wu, Y. (2009). Synthesis and characterization of self-cleaning cotton fabrics modified by TiO_2 through a facile approach. *Surface and Coatings Technology*, 203, 3728–3733, ISSN 0257-8972. https://doi.org/10.1016/j.surfcoat.2009.06.008

98. Ahmad Kamal, S. A., Ritikos, R., and Abdul Rahman, S. (2021). Enhancement of self-cleaning properties and durability of superhydrophobic carbon nitride nanostructures by post-annealing treatment. *Surface and Coatings Technology*, 409, 126912. https://doi.org/10.1016/j.surfcoat.2021.126912

99. McGuinness, N. B., John, H., Kavitha, M. K., Banerjee, S., Dionysiou, D. D., and Pillai, S. C. (2016). Self-cleaning photocatalytic activity. *Materials and Applications. Energy and Environment Series*, 204–235. https://doi.org/10.1039/9781782627104-00204

100. Ibrahim, N. A., Nada, A. A., Eid, B. M., Al-Moghazy, M., Hassabo, A. G., and Abou-Zeid, N. Y. (2018). Nanostructured metal oxides: Synthesis, characterization, and application for multifunctional cotton fabric. *Advances in Natural Sciences: Nanoscience and Nanotechnology*, 9(3), 035014.

101. Sivakumar, A., Murugan, R., Sundaresan, K., and Periyasamy, S. (2013). UV protection and self-cleaning finish for cotton fabric using metal oxide nanoparticles. *Indian Journal of Fibre and Textile Research*, 38, 285–292.

102. El-Khatib, E. M. (2012). Antimicrobial and self-cleaning textiles using nanotechnology, *Research Journal of Textile and Apparel*, 16(3), 156–174. https://doi.org/10.1108/RJTA-16-03-2012-B016

103. Afroz, S., Azady, Md. A. R., Akter, Y., Al Ragib, A., Hasan, Z., Rahaman, Md. S., and Islam, J. M. M. (2021). Self-cleaning textiles: structure, fabrication, and applications, In: Mondal, Md. I. H. (ed.), *Fundamentals of Natural Fibres and Textiles, The Textile Institute Book Series*, Woodhead Publishing, pp. 557–597, ISBN 9780128214831, https://doi.org/10.1016/B978-0-12-821483-1.00016-4

104. Yusuf, M. and Shahid, M. (eds.) (2022). *Emerging Technologies for Textile Coloration*, CRC Press, Boca Raton.

105. Yusuf, M. and Madhu, A. (2022). Smart nanotextiles for filtration. In: Yilmaz, N. D. (ed.), *Smart Nanotextiles: Wearable and Technical Applications*, Scrivener Publishing, USA, pp. 341–360.

Index

abrasion resistance 153, 238, 245
additive 17, 20, 25, 131, 155, 183, 222, 230
adhesion 32, 55, 72, 172, 225, 231, 240
AFM *see* atomic force microscopy
agent 101–102, 180, 244
  acylating 240
  antibacterial 20, 25, 30, 157, 236
  antimicrobial 28
  biological 108
  biological weapon 71
  chemical 74
  flocculating 132
  functionalizing 222
  noise reduction 23
  reducing 174
  silane coupling 231
  therapeutic 28, 33–34, 128
  wetting 132
Alzheimer's disease 181
antibacterial property 21, 151, 172, 233
antimicrobial peptide 8
antioxidant 25, 132
apparel 174, 179, 222
application
  acoustic 22–23
  advanced technology 150
  agricultural 74
  automotive 22
  biological 55
  biomedical 1, 15, 27, 35, 53, 72, 125
  braiding 99
  construction 111
  dental 203
  dielectric 204
  electrospun nanofibers 133
  environmental 47, 76–77, 205
  geotextile 25
  industrial 57, 244
  medicinal 157
  nanofluidic 111
  piezoelectric nanogenerator 109
  scientific 195
  self-cleaning 178, 183
  sensory 106
  synchronous 194
  technical 174
  weaving 99
  wound-healing 14
approach
  chemical 198
  chemical synthesis 196
  colloid suspension 237
  crosslink 235
  electrospinning 2, 127, 129, 154
  finishing 172
  knowledge-based 47
atomic force microscopy (AFM) 241

bacteria 133, 157, 179–180, 222, 235, 237–238
  gram-negative 180
  gram-positive 21
  inflammatory 28
battery 14, 53, 75
bending instability 13–14, 65–67
biocompatibility 1, 27, 30–31, 33, 107
biodegradability 1, 27, 30–31, 111, 150, 176, 200
biomolecule 56, 128–131, 135, 141–142

biopolymer 27, 73
bond 55–56, 178, 200, 227, 237–238, 240
bone tissue 14, 31, 107
bovine serum albumin (BSA) 137–138
Brownian motion 20
BSA *see* bovine serum albumin
bulking process 32
burst release 130–131, 138

carbonization 17, 96, 99
carbon nanofiber (CNF) 19, 22–23, 96, 151
carbon nanotube (CNT) 20, 150–151, 154, 156, 171–172, 221, 232, 236
catalyst 17, 77, 232–233
cell 30, 55, 72–73, 106–107, 133
  animal 182
  cancer 203
  fibroblastic 28
  osteoblast 55
cellulose 19, 23, 25, 27–28, 48, 158, 238
cellulose fiber 24, 157, 231, 239
centrifugal spinning 2, 5, 9, 11, 13, 49, 88–90, 100–103, 108, 111–112
charcoal 71, 74
chemical vapor deposition (CVD) 53, 76, 196–197, 234
chitosan 6, 8, 34, 49, 102, 180
CNF *see* carbon nanofiber
CNT *see* carbon nanotube
coating 18, 34, 76, 153–154, 172–175, 222–223, 225, 228, 231–232, 235, 244
  anticorrosion 195
  continuous 174
  hydrophilic 222, 225
  hydrophobic 222
  nanomaterial 155
  nanoparticle 239
  self-cleaning 222–223, 244
  superhydrophilic 238
  superhydrophobic 229, 231
coaxial nanofiber 75, 134
collagen 27, 49, 106
collector 9, 11, 13–14, 16–17, 57, 59–61, 65, 93–94, 98–99, 103–105, 114
  blade auxiliary electrode 16
  conical 97, 99
  cylindrical 13
  disk 60
  drum roller 60
  electrode-assistant 16
  funnel-shaped 99
  rotating 11, 108
  rotating funnel 98
  spherical concave 89
  static 11, 60
  window 60
composite 22, 48, 52, 76, 111, 198, 241
  hybrid 199
  hybrid materials 198
  metal-oxide 193
  nanofiber-reinforced 76
  nanostructured 150
  non-structural 195
compound 135, 198, 235
  active 132
  bioactive 127, 132
  chemical 129
  fluorine 232
  halamine 222
  moldings 203
  organic 19
  organosilicon 231
  volatile 227
condensation 173, 200, 230
conductivity 10–11, 23, 61–62, 74–75, 109, 137, 172, 199, 229
contact angle 10, 178, 223, 225, 241

contamination 28, 222, 232–233, 236, 239, 242
co-precipitation 196, 198
coronavirus 154
cosmetics 181, 205
cotton 156–158, 228, 236, 238–239
Coulomb repulsion 64
Crohn's disease 181
crystallization 6, 173
CVD *see* chemical vapor deposition

DDS *see* drug delivery system
deformation 49–50, 63, 65
degradation 135, 139, 234
  environmental 183
  thermal 76
deposition 10, 14–15, 20, 30, 51, 53, 76, 173, 196–197, 241
detergent 236, 239, 243
differential scanning calorimetry (DSC) 202, 241
diffusion 20, 22, 129
diffusion coefficient 135
dirt 153, 178, 222, 226, 232–233, 242–243
disease 19, 28, 73, 181
dispersion 135, 180, 235
  colloidal 198
  homogeneous 195
  non-homogeneous 199
dissolution 7, 53–54, 131
distribution 21–22, 101–102, 131, 181
  geometrical 30
  homogeneous 152
  microstructural 206
drainage 25–26
drug 33–34, 73, 108, 125–133, 135, 137–140, 142
  analgesic 126, 135
  antibiotic 74
  anticancer 135
  anti-inflammatory 130
  encapsulated 129–131
  pure 73
  water-insoluble 139
drug delivery 4, 8, 16–18, 27, 33–34, 47, 59, 72–74, 106, 108, 127–128
drug delivery system (DDS) 9, 33–35, 127, 140–142
drug encapsulation 33, 127
drug loading 33, 130–131
drug release 33–34, 70, 73, 77, 108, 126–127, 129–131, 139
dry spinning procedure 63
DSC *see* differential scanning calorimetry

ECM *see* extracellular matrix
electric field 14, 63–64, 68, 98
electric hole 233
electroblowing 111–112
electrocarding 113–114
electro-centrifugal spinning 2, 5, 12–14
electrode 13, 16, 22, 53, 57–58, 60, 63–64, 195
electrospinning 9–15, 17–18, 29–30, 32, 57–58, 60–71, 73–75, 77, 88–92, 99, 102–103, 106–113, 126–127, 129–131, 135–136, 154, 228–229
  air-sealed centrifuge 13
  bath 97, 100
  blend 129–130, 135–136
  bubble 12, 113
  centrifugal 112
  charge injection 12
  coaxial 12, 19, 33, 59, 125–126, 134–136, 235
  concurrent 28
  core/shell 134
  dry 100
  dynamic 106
  dynamic liquid 106

emulsion 18, 127–128, 135–139
hot melt 12
magnetic-assisted 12
multiple jet 12
nanospider 12
near-field 12
needleless 98
one-phase 130
reactive 12
single-needle 59
solution/melt 12
stepped airflow 95, 99
suspension 127, 131–133
thermo-curing 12
three-axial 18
triaxial 5, 12
two-phase 137
electrospinning method 5, 12, 14, 16, 18, 48, 55, 62, 89, 91, 130, 133, 136
electrospinning process 14, 17, 58, 65, 91, 126, 129, 134, 136–137, 228
electrospinning setup 58, 60, 90–91, 95, 98, 110
electrospinning technique 12, 20, 49, 59, 73, 88, 98, 107, 126, 131
electrospun fiber 61, 73, 126, 128–129, 131, 228–229
electrospun jet 61, 64, 66–67
electrospun nanofiber 11, 16–18, 28, 30–31, 33–34, 74–76, 88, 91, 113, 125–126
electrospun nanoyarn 91, 110
electrostatic repulsion 67, 69
emulsion 125–126, 133, 135–139, 142
emulsion nanofiber 135–136
energy 49, 69, 71, 106, 109, 173, 177, 233, 242
kinetic 20, 23
mechanical 109
energy storage 16, 47, 70, 74, 109, 197
environment 28, 70, 73, 132, 151, 159, 170, 176, 181, 183, 245
harsh 130
moist 28
organic solution 129
physiological 35
erosion 129
etching 7, 223, 227
extracellular matrix (ECM) 15, 30–31, 72–73, 107

fabric 151, 155, 169, 171–176, 179, 183, 223, 234–236, 241–244
breathable 74
cellulose acetate 241
cotton 155–156, 159, 179–180, 233–236, 238
cured 235
durable 156
dyed 157
finished 173
flame-retardant 156
functional 150
functionalized 235
nanocrystallite-treated 238
nanoparticle-containing 175
nonwoven 25
nylon 228
protein-based 239
rotating 175
self-cleaning 156, 221–223, 230, 241–243
soil-resistant 226
spacer 23
treated 155, 172
water-repellent 244
fabrication 2, 8–12, 17–18, 31, 33, 35, 57, 99, 110, 141, 243
fiber 4, 13–14, 16–17, 19–21, 27, 48–49, 51, 59, 61–62, 90, 95, 125–126, 129, 133–138, 171–172, 174, 176–177, 180, 227–229, 238–240

abraded 176
acetate 158
bulk 50
coir 25
collagen 31
composite 9
core-sheath 58
drug-loaded 131
electro-centrifugal-spun 13
electrospun photocatalytic 235
fabricated 7
fabricated alginate 9
functionalization of 159, 183
gravity 7
keratin 240
melt-blown 20
nanosized 71
nanoto microscale 14
polyesters 240
porous 16
protein 239–240
spun-bond 19
super-hydrophilic chitosan 21
synthetic 240
textile 48, 173
ultrafine 16, 75
viscose rayon 48
wet 50
film 173, 197, 205, 229
extract-loaded 73
motion picture 48
nanoparticles-based 197
oxide 173
protective 156
solvent-cast 73
thin 173, 197
filter 20–22, 24, 71–72
filtration 4, 14, 16–17, 19–23, 25–27, 47, 70, 72, 87, 229
flame retardancy 151, 155, 159, 195
force 5, 9–10, 13, 15, 56, 61, 64, 68, 70, 154
air shear 112
centrifugal 9, 89, 101–102, 113
Coulomb 66
electrical 11, 64, 67, 138
electrostatic 5, 56, 58–59, 113, 126, 228
gravitational 69
intermolecular 225
mechanical 16, 109, 113
repulsive 13, 65, 67–68
van der Waals 55–56
viscoelastic 64
freeze-drying 6, 32, 53–54
fungi 176, 180, 237

geotextile 19, 25–27, 174
glycosaminoglycans 31, 73
graphene 95, 151, 172

hair dyeing recipe 170
healing 27–28, 32
hybrid method 100, 111–112
hydrolysis 25, 173, 200, 230
hydrophilicity 30–31, 133, 137, 226, 231, 235, 241
hydrophobicity 30, 231, 235–236, 241, 245
hydrophobic polymer 130–131

ibuprofen 34, 140
infection 19, 28, 30, 180

jet 10–11, 13–14, 16–17, 60, 63, 65–66, 68–70
charged polymeric 13, 228
continuous 14
electrically propelled 65
electrospinning 17
electrospun fluid 65
elongate polymer 102
fiber 134
hydrodynamic 68

keratin 27, 49

ketoprofen 140

laser 171, 196, 231
lotus 178–179, 222, 224
lotus leaf effect 179

membrane 21–22, 30, 51–52, 57, 59, 71–75, 183, 233
metal-oxide nanoparticle 195, 200–201, 203, 206
metal-oxide nanostructure 195, 197–199
method
  chemical 7, 57, 172
  core/shell 134
  evaluation 241
  physical 7, 172
  sol-gel 173, 180, 223
  synthetic 172
  traditional 200
  two-step 129
microcapsule 127, 139–141
microorganism 72, 232–233, 236–238
microwave 172, 232, 235
morphology 2, 6, 9, 11, 14, 88, 91, 106, 108–109, 133, 150, 156, 196–199
  bead-less mat 102
  cell 33, 133
  nanometric 228
  noncontinuous 137
  ordered 232
  pore 6
  random 14

nanoclay 150–151
nanocomposite 1, 70, 75, 149–152, 154–158, 160, 169, 193, 195, 199–202, 205
  antimicrobial 157
  drug-loaded 12
  hybrid 203
  metal-oxide 195, 200
  metal-oxide/polymer 200
  polymer-based 154, 194
  polyurethane-based MnO_2-$FeTiO_3$ 155
nanofinishing 149, 151–153, 156
nanomaterial 2–3, 149–153, 155–156, 159, 169–170, 172, 181–182, 202, 221–222
nanorod 3, 56, 59, 197
nanosheet 3, 8, 52, 56
nanostructure 1, 3–4, 11, 18, 52, 56, 150, 193–199, 224
nanotextile 151, 171, 182–183
nanoyarn 88–114
nerve growth factor (NGF) 138
nerve tissue 15, 107
neural stem cell (NSC) 33
neural tissue 15
NGF *see* nerve growth factor
NSC *see* neural stem cell
nuclear magnetic resonance 49
nylon 72, 109, 114, 180

organ 27, 72–73, 181–182
osteoconductivity 31, 107
osteoinductivity 31
oxidation 17, 139, 175, 202, 226

PAN *see* poly(acrylonitrile)
Parkinson's disease 181
particulate matter (PM) 20–21
patient 27, 29, 31, 34, 127, 154
PCL *see* poly(ε-caprolactone)
peptide 8, 56–57
permeability 28, 30, 74, 77
phase separation 2, 5–6, 14, 17, 53–55, 74, 89
photocatalyst 157, 221, 226, 232–234, 244
physical vapor deposition (PVD) 76, 196
PLA *see* poly(lactic acid)
plant extract 28, 157
plasma 197, 227, 240

PLD *see* pulsed laser deposition
PLGA *see* poly(L-lactic-co-glycolic acid)
PM *see* particulate matter
pollutants 19–21, 71, 226, 236
poly(acrylonitrile) (PAN) 10, 14, 17–20, 30, 71–72, 75, 91, 96, 98–99, 109
poly(ε-caprolactone) (PCL) 8, 12, 27, 29–31, 98, 130
poly(lactic acid) (PLA) 20, 27, 76, 103, 130, 204
poly(L-lactic-co-glycolic acid) (PLGA) 31, 77, 133, 138
polyester 23, 174, 234, 238, 240
polyester fiber 228, 236, 240
polymer 1, 4–11, 19–21, 25, 48–49, 60–63, 75–77, 102–103, 130–131, 133, 137, 139–141, 193, 195–196, 200
  biocompatible 130
  biodegradable 72, 129
  cross-linked 61
  fluorinated 231
  gelatin 138
  immiscible 134
  mesostructured conducting 52
  molten 62
  natural 1, 20, 27, 30
  nonbiodegradable 73, 129
  nondegradable 72
  polar 131
  polyurethane-based 73
  synthetic 1, 20, 23, 27, 30, 35
  viscoelastic 7
  viscose 58
  well-electrospinnable 135
polymeric nanofiber 1, 23, 30, 33, 35, 47–78
polymeric solution 13, 49, 51, 62–63, 129, 134, 228
polymerization 49, 57, 200, 227, 230–231
polymer matrix 23, 195, 199, 201–202, 206
polymer solution 6–7, 9, 50–51, 54, 57–58, 62–63, 103, 109, 126, 130–132, 137–138, 140–141
polysaccharide 20, 27, 49, 56
polystyrene 11, 13, 20, 22, 103, 200
polyurethane 20, 30, 49, 72, 202
polyvinyl alcohol (PVA) 12, 20, 22–23, 27, 140
porosity 9, 14, 16, 22–23, 27–28, 30–32, 35, 75, 77, 94, 98
protective clothing 14, 72, 74
protein 20, 27, 56, 126, 132, 203
pulsed laser deposition (PLD) 196–197
PVA *see* polyvinyl alcohol
PVD *see* physical vapor deposition

reactive oxygen species (ROS) 226
reduction 23, 75, 158, 169, 171–172, 174, 177, 180, 183
resistance 71, 75, 153, 156, 159, 176, 179, 183, 199, 202, 235, 239
reverse osmosis (RO) 22
RO *see* reverse osmosis
ROS *see* reactive oxygen species

SBS *see* solution blow spinning
scaffold 7, 30–32, 55, 72, 106
  active 28
  biomimetic 31, 35
  composite 28
  macro-porous 55
  nanofiber-based 55
  nanofibrous 15, 53, 55, 106–108, 139
  nanoyarn 106–107
  tissue 137–138
self-cleaning attribute 233, 236–237

self-cleaning capability 224, 233–234, 241
self-cleaning finish 221–222, 238–239, 243
self-cleaning property 157, 178, 224–226, 228, 232, 234–235, 239
self-cleaning surface 76, 222–223, 229, 231–232, 237, 243–245
semiconductor 7, 226, 233
sensor 4, 15–16, 18, 47, 51, 57, 77, 109–110, 171, 183, 195
SF *see* silk fibroin
silica 12, 157, 172, 176–177, 180, 230, 234–235
silicone 225, 231
silk fibroin (SF) 106–107
silver nanoparticles 20, 95, 99, 154, 157, 170, 174, 180, 221, 232, 236
soil 25, 159–160, 176–177, 179, 182, 222, 226, 242
  fine-grained 25
  reinforced composite 25
  soft-grained 25
sol-gel process 173, 198, 200, 229–231, 234, 239
solution blow spinning (SBS) 10–11, 89, 103–104, 106, 111
solvent evaporation 7, 9–13, 17, 228
spinning process 9, 49–50, 61, 89, 102, 113, 132
stain 179, 233–235, 238, 242
  coffee 235
  color 233
  red wine 234, 239
  tea 244
stress 47, 49, 63, 65–66, 99, 111, 131
substrate 8, 24, 196–197, 227, 231–232, 244
  conductive 16, 33
  heated 197
  polymeric 25
  polyvinyl pyrrolidone 197
supercapacitor 17, 75, 103–105
surface energy 68, 223, 227
surface modification 22, 126–129, 142, 228–229
surface tension 9–11, 61–65, 68, 88, 113, 133, 137, 178, 225, 228, 231
surfactant 56, 132, 136–138
suspension 125–128, 131–132, 142, 173
syringe pump 9, 59, 103
system 33, 56, 99–100, 134–135, 139, 201
  circulatory 183
  computing 149
  concentric nozzle 10
  dust collection 182
  dynamic liquid 98
  dynamic liquid support 107
  electromechanical 13
  implantable 73
  layered fabric 74
  nanofiber collection 103
  non-closed 182
  polymer/solvent/drug 136
  respiratory 182
  self-assembled 55
  self-cleaning functional 223
  two-collector 107

technique 2, 5, 7, 9–13, 48–51, 53–54, 74, 88–91, 100, 102–103, 111, 113, 127–129, 196–198, 200
  cleaning 223
  coating 173
  dip-coating 173
  electrospraying 28
  emulsification 136
  microcapsulation 139
  microwave 235
  nanofiltration 71

one-pot 154
sol-gel 198, 229–230
spin-coating 173
technology 35, 75, 77–78, 113–114, 136–137, 170, 173, 193–196, 221, 225, 229, 235, 242–243
electro-spinning 228
melt blowing 103
microcapsule 141
nanomaterial 151
sol-gel 173
textile 22, 150–151, 153–159, 169–183, 194, 222–223, 227, 229, 231–233, 235–238, 243
antimicrobial 157
automotive 19, 177
coated 231
commercial 182
conductive 171, 183
cotton 155, 159, 234–235
flax 159
medical 19, 27, 157
self-cleaning 157, 223, 226, 232, 241–242, 244–245
smart 109–110
synthetic 237
textile industry 22, 151–153, 156, 170–171, 183
textile substrate 221–222, 234, 236–239, 244–245
thermal conductivity 7, 201, 236
thermally induced phase separation (TIPS) 6–7, 13, 17, 47, 50, 53–55, 61–64, 228
TIPS *see* thermally induced phase separation
tissue engineering 4, 6–7, 16–17, 27, 30, 47, 59, 70, 72–73, 89, 106–108
tissue regeneration 10, 12, 30, 107–108, 139

ultrafiltration 59, 71, 77
UV protection 153, 155, 159, 172–173, 179, 229, 239
UV radiation 155, 179, 241

vacuum 6, 64, 176–177, 196
vapor-induced phase separation (VIPS) 17
vascular endothelial growth factor (VEGF) 138
VEGF *see* vascular endothelial growth factor
VIPS *see* vapor-induced phase separation
viscoelastic model 65
viscosity 9–10, 61–63, 66, 133, 137, 198, 229

wastewater treatment 183, 200
wettability 179, 224–226, 228–229, 241, 244
Wistar rat model 30
wound 12, 27–30, 74, 103
wound dressing 4, 12, 27–29, 70, 72–74, 77
wound healing 28, 30, 73–74

yarn 7, 90–100, 102–104, 107–109, 113–114, 154, 174, 178, 180